Responses to Crime

Responses to Crime

Lord Windlesham

CLARENDON PRESS · OXFORD
1987

Oxford University Press, Walton Street, Oxford OX2 6DP

Oxford New York Toronto
Delhi Bombay Calcutta Madras Karachi
Petaling Jaya Singapore Hong Kong Tokyo
Nairobi Dar es Salaam Cape Town
Melbourne Auckland
and associated companies in
Beirut Berlin Ibadan Nicosia

Oxford is a trade mark of Oxford University Press

British Library Cataloguing in Publication Data
Windlesham, David James George Hennessy, Lord
Responses to crime.
1. Crime and criminals — Great Britain
I. Title
364'.941   HV6947
ISBN 0-19-825583-7

Library of Congress Cataloging in Publication Data
Windlesham, David James George Hennessy, Lord, 1932–
Responses to crime.
Includes index.
1. Criminal justice, Administration of — Great Britain.
2. Crime and criminals — Great Britain. I. Title.
HV9960.G7W56 1987   364'.941   87-12297
ISBN 0-19-825583-7

Set by Cotswold Typesetting Ltd, Cheltenham and Gloucester
Printed in Great Britain
at the University Press, Oxford
by David Stanford
Printer to the University

Law, says the judge as he looks down his nose,
Speaking clearly and most severely,
Law is as I've told you before,
Law is as you know I suppose,
Law is but let me explain it once more,
Law is The Law.

Yet law-abiding scholars write:
Law is neither wrong nor right,
Law is only crimes
Punished by places and by times,
Law is the clothes men wear
Anytime, anywhere,
Law is Good morning and Good night.

Others say, Law is our Fate;
Others say, Law is our State;
Others say, others say
Law is no more,
Law has gone away.

And always the loud angry crowd,
Very angry and very loud,
Law is We,
And always the soft idiot softly Me.

W. H. Auden, *Collected Poems*, 1976

Preface

POLICIES for dealing with crime are constantly in the public eye. Few issues receive so much exposure or evoke such strong feelings. Politicians tend to be judged as much by their instant response, and its tone, as by their considered policies. The policies themselves, if they are to have the effects intended, depend on a realistic appreciation of how the system of criminal justice works. No system can be moulded or adapted effectively unless it is first understood by those doing the moulding and adapting. In England and Wales the administration of justice is complex and difficult to grasp. The practices are often confusing, and the principles, where they can be detected at all, tend to be buried beneath procedural rules. While the political imperative to be seen to be getting to grips with crime is only too evident, there is little common ground as to the directions in which to go. The old certainties of the penal reformers on the one hand, and the upholders of the retributive and deterrent functions of punishment on the other, have been equally shaken. Disillusion is the result, and unrelated policy initiatives in reaction to events the danger.

Criminal justice is in fact comprehensible, with an effort, although it is impossible to rationalize. It has developed over the last century or so, adjusting to public demands as a way—on the whole, a creditable way—of balancing the interests of society against those of the individual citizen whose behaviour has transgressed the requirements of the criminal law. That statement of course is a thumping generality calling for interpretation, which it will receive as the book unfolds. Like much else in our national life, the administration of justice was never designed as a formal system based on order, principle, and fairness. These qualities are to be found embedded somewhere in many of the laws and conventional practices, but they lack visibility and recognition. It is one of the themes of this book that we should not be frightened of rationality and principle, unfashionable as they may be, since they represent the building blocks for the construction of more coherent policies aimed at attracting greater public confidence. As we set out, however, it is necessary to accept that in its present shape criminal justice is less of a system than a largely pragmatic process based upon a sprawling accumulation of experience. The working of each component part will have an effect on other parts,[1] but as the private individual, be he victim, witness, or offender, makes his way through the tortuous and perplexing maze, at least he can cling on to one guide rail: that each person charged with a criminal offence is entitled to be tried and punished according to the circumstances of his particular offence. These circumstances may aggravate or mitigate his culpability, determining the scale of retribution exacted by way of penal sanctions, and bearing on questions of release and the risk of re-offending.

Penal policy today is under mounting pressure, from moderate and sensible people as well as the loud and angry crowd, to ensure the protection of society to the maximum extent possible and to look to the interests of the victims as well as the needs of the offender. Those must be the first aims of criminal justice. But how are they to be pursued? The public mood inclines, increasingly as it seems to me, towards punishment as the main response. Yet the crime, in its enormity or triviality, is already in the past. Is it that act which should be avenged above all else, or should the focus be on the perpetrator: a living creature with a future as well as a past? Each accused person appearing in court is different from any other; what he has done is peculiar to him and his victim. His previous record, his motives, and the chances of his doing it again all need to be investigated and weighed up before the punishment is decided upon. Treating a case on its individual merits is more than simply the due of the offender: it is an essential preliminary to the establishment of guilt, the measurement of punishment, and the prospect of rehabilitation. This is the true meaning of justice; a reconciliation of Auden's We and Me.

To explore these perennial questions in the context of modern penal policy, I have selected certain aspects within my own

[1] For a detailed discussion of interdependence in the criminal justice system, see D. Moxon (ed.), *Managing Criminal Justice*, Home Office Research and Planning Unit, HMSO, London, 1985.

competence. Each chapter has a topical significance, sometimes so much so that the target has been shifting as I have been writing. *Responses to Crime* is not intended as an impartial and comprehensive survey. It does not eschew the personal, the subjective, or the qualitative. What it does is to provide a commentary for practitioners working in the criminal justice system and for interested laymen on some of the ways in which contemporary British society responds to criminal offending. I hope, too, that law students beginning to look beyond the content of the criminal law to its aims and consequences, to who is affected and how, may find some practical guidance and even an occasional insight into how policies are made. As to criminological research, I have done no more than skim the surface of some recent findings. Research surveys and reports are useful primary sources of information about offenders and the manner in which they offend, but the conclusions are too often masked by a dense jargon that is difficult for the non-specialist to penetrate. Greater clarity and simplicity in presentation would encourage a more widespread interest in research on the part of those who have a say in the formulation of policy.

One particular omission should be explained. Although the role of the police in countering crime is mentioned in places, particularly in Chapter 3, it is not the subject of separate discussion since I regard the issues as being too large and my grasp too unsure. Throughout the book, by capitalizing on my own stock of experience, greatly supplemented by what I have learned in writing it, I have sought to be informative and expository, as well as opinionated.

Wherever possible, the statistical references include those for the calendar year 1985. Unless otherwise stated, the source is the criminal or prison statistics which are published annually by the Home Office, or the British Crime Survey. *Responses to Crime* was completed and handed to the publishers at the end of 1986, and thus does not take account of any changes that may result from the enactment of the Criminal Justice Bill, which was introduced by the Home Secretary in the House of Commons on 13 November and received a Second Reading on 27 November 1986. Several of the Bill's proposals are relevant to the book's contents, and I have included references as appropriate in the text and the footnotes. They should be read, however, with an awareness that at present they have no more, or less, than the status of legislative proposals put forward by the Government, to be approved, amended, or rejected during the passage of the Bill through Parliament.

WINDLESHAM

Oxford
December 1986

Acknowledgements

My first acknowledgement is to the Warden and Fellows of All Souls College, Oxford, for electing me as a Visiting Fellow for two terms to facilitate the writing of this book. The companionship of the Fellows and passing guests made the College a most hospitable as well as a scholarly haven. Thanks go too to my colleagues on the Parole Board, whose lively conversation and diverse expertise has been a constant stimulus and deep well of knowledge. Some Board members, past and present, have been generous enough to find the time to read and comment on chapters of the book in draft and to them I am especially grateful. The Library Clerks in the House of Lords and the staff at the Bodleian Law Library in Oxford have been courteous and helpful in pointing me in the right direction and finding the answers to numerous questions.

I also wish to acknowledge a debt to the Home Office. Lord Whitelaw, when Home Secretary, appointed me as chairman of the Parole Board, thus reawakening an interest in penal matters which led in due course to this book. His successor, Leon Brittan, reappointed me for a second term, and the present Home Secretary, Douglas Hurd, has enabled me to spend part of my time in Oxford besides stoically accepting it as unlikely that everything I have written will commend itself to his ministerial ear. Home Office officials, especially David Faulkner, the Deputy Secretary in charge of the Criminal and Statistical Departments and the Research and Planning Unit, have been of much assistance in providing factual information, statistics, and interpretation of trends. It goes without saying that their co-operation implies no agreement with, or sympathy towards, the contents of the book. The impartiality of officials, real as well as proclaimed, leaves them in a less precarious position than the author. While conscious of prevailing departmental orthodoxy, and to an extent subject to it, I am neither an apologist for current policies towards crime nor an outright critic of them.

This in-between stance leaves me exposed to cross-fire from several quarters. What may strike root-and-branch radicals as hopelessly reactionary and complacent may be regarded as speculative and ill-advised by those who have the awesome responsibility of managing the process of criminal justice. All I can say in reply is that this is how the contours of the penal landscape seemed to me from my own particular vantage point, a perspective determined by the scrutiny of large numbers of cases of individual offenders.

As the book progressed, I had an opportunity to rehearse some of the ideas it contains on various platforms, including a broadcast lecture on BBC Radio 3, the 1985 Annual Conference of the Howard League for Penal Reform, and a seminar held by the Centre for Criminological Research in Oxford. The head of the Centre, Dr Roger Hood, University Reader in Criminology and a Fellow of All Souls, has been an unfailing source of encouragement and guidance from the start. Dr Joanna Shapland, a Research Fellow at the Centre for Criminological Research, read the manuscript, correcting errors and making valuable suggestions for improvements. I also benefited from, and greatly enjoyed, an Anglo-American conference on penal policy held at Ditchley Park in June 1986, and have exercised my prerogative as conference chairman to draw on some of the comparisons made between our two systems. As time has gone on I have resisted the temptation to publish any of the chapters as separate papers or articles, even in abbreviated form, although in doing so I have had to endure moments of frustration as swifter hares passed by.

None of those who have helped by making comments on various chapters should be identified in any way with the book's contents or orientation. In addition to those mentioned above, I wish to thank the following for providing information and comments: Lord Allen of Abbeydale, Dr Andrew Ashworth, Dr George Birdwood, Sir James Crane, John Dellow, Dr David Farrington, Professor Abraham Goldstein, Lord Hutchinson of Lullington, Dr Terence Kay, Gilbert Kelland, Sir Anthony Lloyd, the National Association for the Care and Resettlement of Offenders, Helen Reeves and the National Association of Victims Support Schemes, Dr Julian Roberts, Dr Donald Scott, Michael Smith and the Vera Institute of Justice, Eric Stockdale,

Dr Anthony Storr, Clifford Swann, Professor Nigel Walker, Dr Barbara Williams and the Rand Corporation.

Particular thanks are due to Dr Valerie Johnston for her meticulous work in checking the statistics and references; to Mrs Audrey Knapp for long and patient hours spent at the word processor; and to Richard Hart, Law Editor of the Oxford University Press.

The extract from 'Law Like Love' from *Collected Poems* by W. H. Auden (edited by Edward Mendelson) is reprinted by permission of Faber and Faber, Ltd and Random House, Inc., New York, and the quotation from an article by A. S. Goldstein by permission of the Editor, *Law and Contemporary Problems*.

Contents

List of Tables

1

The Setting: Within the Toleration
of Public Opinion

I

Visualize, if you will, a man convicted of a criminal offence, with his two associates, uneasily awaiting sentence. All three men are in their early twenties and each of them has been in court before. Their manner reflects a shifting mixture of bravado and apprehension. On the bench, raised on a dais, sits the judge, sometimes a High Court judge, but more likely a full-time circuit judge or an experienced barrister or solicitor sitting as a part-time recorder. For this is a Crown Court with the power to try the more serious criminal offences on indictment.[1] The defendants have admitted to charges of robbery and burglary, taking and driving away a car without the owner's consent, and driving without licence and insurance. Since each of them has pleaded guilty, there is no jury.

No room for doubt exists about the facts of the robbery offence. The men had been caught in the act by the police as they ran out of a newsagent's corner shop, having stolen the contents of the till amounting to a little more than £180 in cash. The first named defendant had used a knife with an 8-inch blade to threaten the shopkeeper, a middle-aged Asian woman, while the second defendant held down her twelve-year-old son who was helping his mother in the shop. The third man was the driver of the stolen car. He remained outside, an unobservant and ineffective lookout, having failed to notice the arrival of the police until too late. Unknown to them, the shopkeeper, worried by stories she had heard of robberies at other premises in the locality, only

[1] Crimes are classified for procedural purposes according to the method of trial. An offence triable on indictment is known as an indictable offence and the trial, always with a jury if the defendant pleads not guilty, takes place in the Crown Court. Summary offences are tried in the Magistrates' court without a jury. Offences triable either way may be tried on indictment in the Crown Court or summarily in the Magistrates' court.

a few months earlier had consulted the crime prevention officer and decided to fit an alarm system. This had been activated when she saw the men outside her shop.

On being arrested and taken to the police station, the defendants readily admitted to a number of burglaries (not all of which involved the same men), in addition to the more serious crime of robbery. One burglary charge only was pursued; various others were taken into consideration.[2] Before hearing what defence counsel could say in their favour, the judge turned to the social enquiry reports prepared by the probation service. These reports are key documents describing the family background and domestic and personal circumstances of those appearing before the court. Details of previous offences and response to supervision are separately listed, and the picture portrayed was depressingly familiar. The first defendant, said to have been the initiator of the robbery, had the worst record, having been convicted on ten previous occasions. Yet his personal history was distinguishable only in degree from the other two in its poverty, neglect, and deprivation. His list of previous convictions included assault, burglary, possessing an offensive weapon, and numerous offences involving motor vehicles. He had undergone Borstal training (since replaced by youth custody), and a shorter period at a detention centre before receiving his first adult sentence of twelve months' imprisonment.

Apart from these custodial penalties, fines and conditional discharges had been incurred; sentences of imprisonment had been suspended and sometimes breached; probation orders and compulsory community service had all been tried. The sheer diversity of non-custodial alternatives to imprisonment was

[2] A defendant pleading guilty to specific charges may also admit to having committed other offences. These are listed on a form prepared for the court which the defendant signs before sentence. Offences taken into consideration should not be more serious than the substantive charge or charges, or markedly different in character, although they may be numerous. The court may increase the severity of the sentence in view of the admission, but may not exceed the maximum penalty in respect of the offences for which the defendant is convicted. The convention is that a defendant who admits his guilt in this way will not subsequently be prosecuted for the offences taken into consideration. Burglary is a lesser crime than robbery, involving entry to a building as a trespasser with an intent to steal or commit certain other offences. Robbery is theft aggravated by the use of force or the threat of force. Cross and Jones, *Introduction to Criminal Law* (10th edn), ed. R. Card, Butterworth's, London, 1984, p. 242.

exemplified in the record of this single offender. The probation officer, who had known him over the years, was unable to hold out much hope. There was nothing to show, he said, that the penalties imposed by the courts, custodial or non-custodial, had resulted in any noticeable disposition to co-operate with the probation service. Nor had any inclination been evident on the part of the offender to detach himself from his friends and associates who formed a deviant peer group, or to try and make a break with a pattern of consistently lawless behaviour.

Studying the details of this dismal record in the papers before him, and gazing at the offender, the judge's mood closely matched that of the man in the dock. Both were resigned rather than despairing, for both had been here before. Now it was up to the judge to decide on the sentence. Had there been relevant medical evidence, as there might well have been in the case of a drug addict or any offender who had received psychiatric treatment, it would normally be considered at this point. In this case there was none.

In the light of the gravity of the robbery and its impact on the victim and her son, as well as the repercussions on the local community, all of which counted for more than the sum of money stolen, everyone in the court sensed that a custodial sentence was highly probable. The length, however, would reflect the judge's view of the seriousness of the offence and the defendant's past record. The judge was not alone in pondering, as he often did, on how it was that a relatively young man, still in his early twenties, could have accumulated such an extensive record of previous offences. But the social enquiry report, in its neutral and dispassionate language, provided the answer.

It described the birth of an unwanted baby, twenty-three years before, to a young woman living in an urban area of high social deprivation. The child's mother was co-habiting at the time but the infant never knew his father. Later his mother married, impetuously and unsuccessfully, and relations between her son and his step-father were never close. Violence and aggression were prevalent within the home, alcoholism was seldom far away, and material standards were low. The housing estate where the barely adequate mother struggled to bring up her three children by two different fathers was decaying, vandalism was rampant, and delinquency was a way of life for many adolescents.

The social enquiry report referred to a pattern of disruptive behaviour emerging early, from about the age of ten. Following a first appearance in the juvenile court when he was twelve, the boy had been made subject to a care order. After psychological tests, he was placed in a special school for maladjusted children where he remained for two years. Intermittently he resided at a local authority children's home, punctuated with periodic returns to his mother for a few months. Further offences of delinquency followed, resulting in a sentence to a detention centre for three months. This appeared to have a salutary effect for a time, and he did not offend again for nearly a year. Boredom, adolescence, and group pressures to conform to the cult standards of delinquent youths on the housing estate were too strong, however, leading in due course to incidents of theft and an assault on a police officer. This time his offending was punished by a period of Borstal training. As the report laconically observed, in an area where a criminal conviction is as good as an old school tie, the young man had now graduated.

Released after nine months, with some improvement in his below-average standards of literacy, he went back to mother, but back also to the dangerous company of his delinquent friends and associates. Once again, hopes that he was beginning to settle down and keep out of trouble were dashed, notably by an emergent pattern of heavy drinking, coinciding with an obsessive attraction towards cars and motorcycles. An enterprising probation officer had managed to find an opportunity under the terms of a supervision order for the offender to attend a motor training project working as a paint-sprayer in a garage, but this employment came to an end when our man breached his probation order by taking a car from the garage without the consent of its owner or the manager of the garage. The offence was compounded by the fact he had neither a driving licence nor an insurance certificate. Once in trouble and bailed to appear in the Crown Court, he had embarked on a spree of thefts and burglary which, taken with the stolen car, led to a prison sentence totalling twelve months.

By the time of the robbery at the corner shop, there was no glimmer of evidence to suggest that the punishment endured had deterred, or that the distress and disappointment caused to relatives and law-abiding friends had negated, the disposition to offend. The robbery itself was planned casually in the course of drinking in a public house as a way of getting some money from a soft target. The defendants had not wanted to hurt anyone, they claimed, but took the knife along for their own protection. The Asian victim's business had not been pre-selected, and any racial motivation was denied. All they were looking for was a shop or a sub-post office in a quiet locality of the city where they could obtain some money easily and get away without being detected. When interviewed by the probation officer for the purpose of reporting to the court, none of the men expressed remorse for the victims, only regret that they had been caught and were now facing the prospect of a prison sentence. The probation officer concluded his reports on each of the men by saying that he recognized that the court would be considering a custodial sentence in view of the gravity of the offence, and that the defendants were well aware of this. In the circumstances, the probation service was unable to offer any alternative recommendation.

Having digested the contents of these reports, and heard counsel in mitigation, the judge asked the three men in the dock to stand up. They had, he said, come to the court and pleaded guilty to the offence of robbery. They had also pleaded guilty to other charges, but those were less serious by comparison. Each of the defendants had behaved in an appalling way. He was not at all impressed by their explanation of how they came to be in possession of the knife. Shopkeepers, he declared, of all colours, creeds, and denominations, worked hard to supply a public service, and those people who thought that small shops and shopkeepers were an easy touch had got to learn they were wrong. He went on:

They have got to learn that shopkeepers, just like other members of the public, are entitled to look to and will get protection from the courts. Any sentence which I impose upon you today, after giving credit where it is due, will be designed to deter you from doing this sort of thing again; but, equally important, to deter other people from behaving in this way and to let the public—and shopkeepers in particular—see that they are going to be protected.

I have listened to what counsel has been able to say on behalf of each of you, and he has managed to find quite a lot to say. One of the things I take into account, and give each of you credit for, is the fact not only that you have come here and pleaded guilty, but, once you realized that the police had caught up with your offending, you were frank and put your cards on the table. Two of you have asked me to take certain other matters into consideration, which shows an element of contrition, as though you want to clear up the whole of your background and perhaps start afresh. I earnestly hope you do.

The sentence I impose upon you must be one involving loss of liberty, but for the reasons mentioned I am reducing the sentence to less than what it might otherwise have been.

Addressing the first defendant, the judge then said that for the main robbery he would have to serve a sentence of four years' imprisonment, with lesser terms concurrent on the other charges. The two associates were sentenced to a total of three years' imprisonment each.

II

It may be thought that this stark episode in its unremitting bleakness is exaggerated and untypical. Statistically, it represents a more extreme case, since offences of robbery account for no more than 1% of all convictions for indictable offences. But convictions result only in a minority of cases reported to the police, with other robberies going unreported. Over the decade between 1975 and 1985, crimes of robbery became increasingly common, growing from 11,300 to 27,500 incidents recorded by the police in the course of each year.[3] By September 1986 the total for the previous twelve months had reached 30,000.[4] While the chances of becoming a victim of robbery may not be high, nevertheless, 30,000 is far too substantial a figure to brush aside. As in the later chapters where case histories are cited, I believe we can learn as much or more about the nature of offending, and about the characteristics of offenders, from the contemplation of narrative accounts as from the scrutiny of official statistics.

---

[3] The number of recorded offences of robbery in 1985 was 10% higher than in 1984, which in turn was 13% higher than in the previous year. The average annual increase between 1975 and 1985 was over 9% per annum, giving a total for 1985 nearly two-and-a-half times that of ten years earlier. *Criminal Statistics England and Wales 1985* (Cm. 10), HMSO, London, 1986, p. 22.

[4] *Home Office Statistical Bulletin* 39/86, 12 Dec. 1986. The increase over the twelve months from Oct. 1985 to Sept. 1986 amounted to 14%.

Plenty of figures will be found in the tables that follow, illustrating trends as well as indicating the numbers involved. They are supplemented and brought to life by the case histories, which are not necessarily typical, but are worth thinking about. In order to preserve confidentiality, names and other identifying features have been deleted, although each element in the narratives is based on fact. Thus, the sequence of events set out above will be entirely familiar to anyone working in the criminal courts or with offenders. Not all of the factors pointing so ineluctably towards crime will always be present, but the circumstances described are quite ordinary.

The student of penal policy needs to probe behind the depressing recital of crimes and sentences to find out what he can about the upbringing and personality of the offender. No stereotype can convey a young man's hopes and fears in adolescence, or the impact on him of constant verbal aggression and physical violence in the home; of his rejection by the father he had never known and the inadequacy of his mother; of his resentment at what he regarded as the greater affection shown towards his younger brother and sister. In their way, the various institutions to which he was sent provided a temporary respite of security and stability which was so notably lacking in the home. Efforts were made in them to develop his character and to remedy his neglected education—at seventeen, he could still only read and write with the facility of a ten-year-old.

Against these more positive features was the fact that some of the fellow trainees at the detention centre and Borstal were an indisputably bad influence. Frequently the talk consisted of little more than a litany of loud-mouthed boasting of what further misdeeds were contemplated on release. To many (although not all) trainees, authority in its various forms was looked on with suspicion, verging in the case of the police towards outright hostility. It was not unusual for the hard-pressed and well-intentioned probation officers with the responsibility for supervising offenders in the community to be mocked and derided, being regarded less as friends and mentors than as latent adversaries capable of returning offenders to the jurisdiction of the courts. Lack of education and maturity, combined with few training and employment opportunities (and with those that are available often being rejected), mean that many persistent offenders have never experienced the routine and incentives of regular work. Behind it all lies a habitual lack of self-esteem, blighting personal relationships and contributing directly to a lack of esteem and respect for other people or their property. A crippling sense of failure is deeply engrained in many younger persistent offenders; and it should always be high among the aims of those concerned with the treatment of offenders to provide targets, however modest, which are within reach of accomplishment.

There is no simple answer to the question of who commits crime, or why they do it. It has not proved possible for even those closest to the penal system to forecast with any degree of accuracy levels of criminality, or convictions in the courts, or fluctuations in the prison population. Opinions abound, not always supported by such impartial information as is available— in particular, the extensive British Crime Survey (BCS).[5] Although such surveys can produce few if any definitive answers, they can delineate the parameters of the probable and the less probable, indicating trends and casting light on the search for truth. The experience of those who come into first-hand contact with offenders, including probation officers, the police, and the medical profession, contributes towards a deeper knowledge of the mentality and personality of offenders as well as the under-lying causes that lead to criminality.

All of this, of course, is generalization, and generalization is the enemy of rationality. The greatest virtue of the English penal system is to be found in the way it has avoided generalizing, preferring to concentrate on the particular circumstances of each offence and each offender. Close acquaintance with the records

[5] The British Crime Survey (BCS) is a large sample survey conducted on behalf of the Home Office in some 11,000 households in England and Wales. Two surveys have so far been carried out, in 1982 and 1984. The survey reports estimate the extent of various types of crime and differ from police statistics in that they include incidents that have gone unreported to, or unrecorded by, the police. The main findings have been summarized in a series of government publications: M. Hough and P. Mayhew, *The British Crime Survey: First report* (Home Office Research Study no. 76), HMSO, London, 1983; P. Southgate and P. Ekblom, *Contacts between Police and Public: Findings from the British Crime Survey* (Home Office Research Study no. 77), HMSO, London, 1984; M. G. Maxfield, *Fear of Crime in England and Wales* (Home Office Research Study no. 78), HMSO, London, 1984; M. Hough and P. Mayhew, *Taking Account of Crime: Key Findings from the 1984 British Crime Survey* (Home Office Research Study no. 85), HMSO, London, 1985.

of a large number of offenders never ceases to startle with the unexpected strengths and weaknesses of human nature. Sometimes the most crushing adversities can be overcome; old lags do change their ways; affection, example, and perseverance can inspire reform and rehabilitation. Against this, on the dark side, there is brutality and calculated violence; cupidity and deceit on a monumental scale; sexual excesses without thought for the victim; and the self-destructive consequences of addiction, whether the agent be drugs, alcohol, or gambling. The most important single prerequisite for a more informed and efficacious penal policy is the recognition that understanding the mind and circumstances of the offender does not imply condoning the crime or putting the interests of the criminal before those of the victim. Like any other social phenomenon, there is no prospect of crime being countered effectively, still less reduced in incidence, unless it is understood.

First of all is the significant fact that only a small minority of offenders are engaged in organized, deliberate crime. That said, professional criminals are capable of inflicting infinite damage on society, and those who make their living off crime are the subject of Chapter 3. Many more offenders are aimless recidivists of the kind portrayed above. Another large category, spread between the prisons, the special hospitals, and other institutions for the mentally disordered, are those who have committed crimes, sometimes serious crimes of violence to the person, while their state of mind has been temporarily or permanently disturbed. The dividing line between remedial treatment and penal responses in the cases of mentally disordered offenders is notoriously hard to identify, and the questions that arise are pursued in Chapter 4. Youthful delinquents are responsible for a huge volume of property offences (the peak age for offending is fifteen), in complete contrast with the person of good character who commits one crime in a lifetime, sometimes minor but sometimes very grave indeed in its consequences, such as the motorist who kills a pedestrian or another road user when he is driving under the influence of excess alcohol.

Malevolent and extensive as the abuse of alcohol is in its contribution towards the commission of crime, it is overshadowed by drugs. This is not because the total volume of crime resulting from drug-taking is necessarily on a larger scale—

indeed, the probability is that it is at a lower level than alcohol-related offending. But the fact that the unauthorized possession and supply of controlled drugs are themselves criminal offences has created a vast service industry, aptly compared by Douglas Hurd to a modern equivalent of the slave trade,[6] which feeds and profits from the needs of addicts. Many drug-takers also resort to theft or other crimes to obtain supplies or finance their habit. Thus, the secondary effects in terms of law-breaking are boundless, although they are not always evident at the time of trial and conviction. These issues, and the public policies adopted in response, are explored in Chapter 6.

Moving up to the topmost peak of crime, offences of homicide, it is remarkable to find that so many murderers are first-time offenders who already knew the victim. In no less than three-quarters of all murder convictions, the victim and the offender had an existing relationship, whether living under the same roof or knowing one another less intimately. Yet however great the personal friction and its causes, and whatever the extent of the frustration, provocation, or misplaced compassion which may have preceded the fatal deed, the fact remains that a human life has been prematurely terminated. It is not by chance that the crime of murder, and the penalty for murder, have long been the most contentious of all penal issues. They are discussed in Chapter 5.

Before considering in detail each of these disparate types of offender, and the ways in which policies to protect the public from the consequences of their criminal acts should be framed, we should pause to note some research indicating that a majority of crimes never come to light at all. No one can be sure of the extent of unreported crime, but findings from the 1984 British Crime Survey show that, of the crimes surveyed, almost two-thirds were not reported to the police, whereas, of those crimes that were reported and recorded, some two-thirds were not cleared up.[7] Many of these unreported incidents will have taken place in the home, or at the work-place, or at the public house or social club. Some will be the result of disputes to which both parties may have contributed; while offences involving serious personal injury such as wife-beating, child abuse, or acts of

---

[6] *The Times*, 23 Sept. 1986.
[7] Hough and Mayhew, *Taking Account of Crime*, p. 16.

sexual assault may have deeper psychological causes. Never the less, the sheer scale of unreported offences is alarming, particularly as not all types of crime were covered by the survey.

For the purposes of comparing reported and unreported crime, the survey omitted the theft of milk bottles from the doorstep, which emerged as the most common, if least serious, crime in the country. According to these findings, stealing milk has become almost a national pastime, with one in ten of all households having had milk stolen from the doorstep on one or more occasions over a period of a year. The survey also excluded crimes without identifiable individual victims. Consequently, no account was taken of quite significant offences like shoplifting, fraud, and vandalism to public or commercial property, or of what are designated as 'victimless' personal crimes such as the possession of illegal drugs.

When thinking about, or speaking about, crime and what can be done to counter it, it is necessary to keep in mind the fact that reported crime, the source of all official statistics, represents only a part of the unquantifiable volume of total offending. While offences involving violence were heavily outweighed by sometimes trivial offences involving theft and damage to property, the conclusion of the Crime Survey report was that most unreported crimes were comparatively minor, but an important minority were not.[8] Equally, it is true to add that, while some of the offenders, especially those responsible for more serious unreported incidents, may have been brought to justice for other offences, an important minority will not be.

III

The continuity of the proceedings in court, the pomp and majesty of the law apparently so little changed in its outward manifestations since Victorian times or earlier, obscures an evolutionary process that in its slow and inconspicuous way is capable of introducing some far-reaching and lasting changes. The scope of the criminal law—what is or is not regarded as illegal by Parliament and the courts—similarly responds to shifts in the public mood. No longer, for example, are homosexual acts

[8] Ibid., pp. 20-1, 51.

between consenting male adults carried out in private treated as criminal offences.[9] Since 1961 the same has applied to attempted suicide. Conversely, the possession and supply of controlled drugs are punished severely, while since 1985 it has been a criminal offence to sell or be in possession of alcohol at football grounds during an event.[10] At the end of the same year, after a long, drawn-out review of the basis on which public order law depended,[11] a Bill replacing the old common law offences of riot, unlawful assembly, and affray with statutory offences was introduced in the House of Commons. A new offence of violent disorder was created, together with additional protection against acts of hooliganism that cause harassment, alarm, or distress. These provisions were enacted in the Public Order Act 1986. New sanctions also take shape, a notable instance being the power of the courts to order confiscation of the proceeds of crime in certain cases.[12]

Procedure in the courts evolves gradually—sometimes, it seems, almost imperceptibly. But some changes are visible to all. In a notable departure from the hallowed tradition of unanimity, majority verdicts were allowed by the Criminal Justice Act 1967 if the jury is unable to agree, partly because of the menace of professional criminals attempting to bribe or threaten individual jurors. Another important innovation is the transfer of the role of the prosecutor in trials for criminal offences away from the police to an independent Crown Prosecution Service, thus separating the decisions on prosecution from the investigatory work of the police. Among the advantages sought from the introduction of an independent prosecution service is the establish-

[9] Sodomy was made illegal as long ago as 1533 and originally carried the death penalty. Other homosexual acts between males in the nature of gross indecency were criminalized only in the nineteenth century.

[10] The Sporting Events (Control of Alcohol) Act 1985 renders illegal both the possession and the sale of alcohol in sports grounds during an event, or in the course of entry to an event. It also makes it an offence to be drunk in a sports ground during events or in the course of entry.

[11] A review of public order law was announced by William Whitelaw when Home Secretary in June 1979. A Green Paper followed in June 1980, and in the same year the House of Commons Select Committee on Home Affairs published a report on the subject. In 1981 Lord Scarman reported on the Brixton riots, and in 1983 the Law Commission recommended reforms in the law relating to riot, unlawful assembly, and affray.

[12] The powers relating to confiscation in the Drug Trafficking Offences Act 1986, and the Criminal Justice Bill 1986/7, are discussed in Chapter 3.

ment of common standards of consistency, selectivity, and professionalism in prosecution practice.

In the case cited as an introduction to this chapter, the essentials of the system of criminal justice as it has grown up in England and Wales stand out in clear relief. In the Crown Court the principal actors are the prosecuting counsel, the accused person, the defence counsel, the witnesses, the jury (if guilt is denied), and the judge. For lesser offences, lay magistrates combine the functions of judge and jury. Rules of evidence and other rules of procedure safeguard the individual citizen charged with the commission of a criminal offence and it is for the prosecution to prove the case. While the adversarial system has its critics (we shall come back to this in Chapter 4), the crucial issue of establishing guilt or innocence beyond reasonable doubt is decided not by the judge in the Crown Court, but by a jury of twelve men and women chosen at random. Each defendant has the right to object to up to three jurors without giving any reason. This right of peremptory challenge has been questioned on the grounds that, in multi-defendant cases especially, where challenges can be pooled, the balance of a jury can be distorted by the removal of those whose appearance and demeanour suggests a respect for the law, or an understanding of its purposes, which may be inimical to the interests of the defence. The Government proposes the abolition of the right of peremptory challenge in the 1986/7 Criminal Justice Bill, while retaining the right to challenge with cause. In order to inform the debate, a survey was carried out of the use that had been made of peremptory challenges in a sample of Crown Court cases during 1986. This indicated that jurors had been objected to on average in 63% of cases in which the Director of Public Prosecutions was responsible for prosecuting in inner London. In multi-defendant trials in the same category, one or more peremptory challenges were used in 75% of cases. Outside inner London and in non-DDP cases, the rates of challenge were considerably lower.[13]

---

[13] Preliminary results of the survey on peremptory challenge and the Crown right of stand-by in cases prosecuted by the Director of Public Prosecutions or the Crown Prosecution Service were made available by the Home Office in January 1987 while the Criminal Justice Bill was in Committee in the House of Commons.

Apart from murder convictions, where a mandatory penalty of life imprisonment is prescribed by Act of Parliament, the courts have the power within a range of sentencing measures to choose the one that seems best suited to the requirements of a particular case in the light of the available information about the circumstances of the offence and what is known about the offender. Maximum penalties are laid down by statute, and judges are expected to follow any relevant sentencing guidance promulgated by the Lord Chief Justice and the Criminal Division of the Court of Appeal. The canvas of this book does not extend to Scotland, with its distinctive legal system, or to Northern Ireland, where there are certain differences, although in both jurisdictions the penalties imposed by the courts once an offender is convicted, and the type of penal institution to which he may be sent, are broadly alike throughout the United Kingdom.

Some critics despair at what they regard as the unyielding quality of the English system of criminal justice, which they argue leads to unimaginative and unnecessarily harsh sentencing of offenders, with too ready a recourse to imprisonment. In sharp contradiction, other voices are raised in protest whenever a judge, having heard all the details of a case, particularly if it involves a well-reported and scandalous incident which strikes a responsive chord in the minds of those who write for the editorial columns of the popular press, prefers to suspend a sentence or put an offender on probation rather than send him to immediate imprisonment.

It is, of course, impossible to satisfy everybody, or even the main currents of opinion when they conflict, as they often do. What has to be done is to seek to build public confidence in the process of criminal justice, while at the same time ensuring that each individual offender is treated in accordance with the requirements of natural justice in that he receives punishment comparable with others who have committed similar crimes. The reference to 'natural justice' calls for a word of explanation. While lip service is readily paid to the need to enforce the rule of law, and to do so impartially and without political interference, the equivalent need to act fairly and reasonably is less obvious. Yet, unless the provisions of the criminal law and their application by the courts are buttressed by generally accepted principles of natural justice, the rule of law will always be at risk of being undermined. In any free society, the strength of allegiance to the legal system will depend on how successfully enforcement of the law is harmonized with the values that underlie it.

Absolute consistency in the sentencing of offenders may be beyond reach, but it is an ideal worth aiming at. Currently, the main instrument is the right of a defendant convicted and sentenced in the Crown Court to seek leave from the Criminal Division of the Court of Appeal to appeal against the conviction, or the sentence, or both.[14] If leave to appeal against sentence is granted, the Court of Appeal has power to reduce the sentence so far as it regards it as excessive or wrong in principle. The right to appeal against sentence, which is not generally available to either the defence or the prosecution in the US federal courts, or in a majority of state jurisdictions,[15] acts as a regulator to the idiosyncrasies of sentencing judges, ironing out inconsistencies between them; and its utilization is fundamental to defensible sentencing policies. Appellate review of sentencing, now established for nearly eighty years, has led to a growing body of case law, supplemented more recently by a small number of guideline judgments in the Criminal Division of the Court of Appeal.

Guideline judgments are restricted in scope, depending on the cases coming before the Court of Appeal and the disposition of the court to influence sentencing practice. While they are reported in the law reports and sometimes in the quality press, and are promulgated to sentencers at seminars arranged by the Judicial Studies Board,[16] they deserve a wider public. There is a

[14] The Court of Criminal Appeal was originally set up by Act of Parliament in 1907, largely to provide a way for the courts to correct miscarriages of justice resulting from wrongful convictions. The right to appeal against sentence was seen as subsidiary to the main purpose of the legislation.

[15] American practice as it was in the early 1970s is described (and strongly criticized) by a federal judge, Marvin E. Frankel, in *Criminal Sentences: Law without Order*, Hill and Wang, New York, 1973.

[16] The Judicial Studies Board was first set up in 1979 following the recommendations of a Working Party on Judicial Studies and Information under the chairmanship of Lord Justice Bridge, as he then was. Its terms of reference were 'To determine the principles on which judicial studies should be planned, to approve the proposed forms of study programmes, to observe them in operation and to report on them periodically to the Lord Chancellor, the Lord Chief Justice and the Home Secretary.' Members are appointed by the Lord Chancellor, in consultation with the Home Secretary and the Lord Chief Justice. For the first six years of its existence the Board exercised its functions only in relation to the criminal jurisdiction of the higher courts; but in 1985 its role was expanded to cover judicial training also in the civil and family jurisdictions, and supervision of the training of magistrates, judicial chairmen, and members of tribunals. A Lord Justice of Appeal, Sir Michael Mustill, has been chairman since 1985.

high degree of interest in the sentencing of offenders, much of it centred on a small but troublesome number of cases attracting partial and exaggerated media coverage. Anything that would divert at least some part of this interest towards the general principles and practices underlying the imposition of sentences would be a move in the right direction. This was well demonstrated when the Court of Appeal laid down guidelines on sentencing in cases of affray. The appellants were three young men who had been convicted for their part in the Tottenham riots in October 1985 at the Broadwater Farm Estate in north London during which a police officer was killed. Over a year later, when the Court of Appeal came to consider the sentences of seven and five years' imprisonment or youth custody which had been imposed in separate trials, public feelings of outrage had cooled. Setting the offences strictly in the context of affray, the Lord Chief Justice, in giving the judgment of the court, reduced the sentences to four-and-a-half and three-and-a-half years, respectively.[17]

IV

Some penal reformers would like to go further than giving more publicity to guideline judgments, and provide a link between the interested public and the judiciary by the establishment of a sentencing commission composed of non-judicial members as well as senior judges. There are precedents in the United States (although the discrepancies there are far greater than anything experienced in the United Kingdom), and the theoretical arguments for a broadening of the interests represented in framing sentencing policies are quite attractive.[18]

The main objection is that this path would lead to an ill-defined hybrid lying half-way between Parliament, which decides on the maximum penalties, and an independent judiciary, which decides on how they should be applied in individual cases. The distinction is less clear-cut in practice, and the evolution of

---

[17] *R. v. Keys and Others*, Law Report in *The Times*, 22 Nov. 1986.

[18] The case for a Sentencing Council is developed by A. Ashworth in *Sentencing and Penal Policy*, Weidenfeld & Nicolson, London, 1983, pp. 447–50.

guideline judgments by the Court of Appeal is a cautious step out on to a perilous no man's land lying between two of the estates of the realm. At this stage, a sentencing commission is still looked on with judicial suspicion, but it is a development that may come more by stealth than by frontal assault. One of the arguments successfully deployed against a proposal canvassed by the Home Office in 1986,[19] to place on the Judicial Studies Board a statutory obligation to publish and disseminate to judges and others the sentencing guidance given by the Court of Appeal, was that it could turn out to be a Trojan horse. Most publishers like to feel they have some influence on what they publish, and there was a possibility that the Board might want to push out the boundaries beyond codification and publication, suggesting to the Court of Appeal fresh areas in which guidelines would be useful. Some judges feared that the door might be opened in this way to external interference in the sentencing process.

Since criminal appeal rights were first introduced in 1907, only the defence has had the right to appeal against sentence although until 1966 the Court of Appeal could increase as well as reduce the length of a sentence on appeal. In the Prosecution of Offences Bill in the 1984/5 Parliamentary session, an attempt was made by the Government to enable the Attorney General to refer to the Court of Appeal any case where he considered that there were grounds for regarding the sentence as excessively lenient. Although the defendant could not have his sentence increased, the Court of Appeal would be given an early opportunity to review and pronounce upon what the proper penalty should have been. In this way, it was argued, public confidence in the administration of justice would not be damaged by what might appear to be an inexcusably lenient sentence passed on a very serious offender, while the actual defendant would be saved from being placed in jeopardy once again by the proviso that his sentence could not be altered. In the desire to be fair, this compromise satisfied almost no one. The Law Officers were apprehensive that the effect of such a change would be to put the Attorney General in a difficult if not impossible position in responding, or not responding, to public

---

[19] *Criminal Justice: Plans for Legislation*, p. 7.

clamour or political pressures. The House of Lords listened to eminent judicial voices endorsing the doubts expressed by lawyers and others: four past or present law lords joining one ex-Lord Chancellor and one former Attorney General in speaking against the clause in committee. A spirited defence by Lord Hailsham from the Woolsack failed to convince the House that the prosecution should be entitled to intervene in the sentencing process, even in the limited way proposed. After a division, the clause was omitted from the Bill by a majority of 140 votes to 98.[20] It was not reintroduced when the Bill went to the Commons.

In the 1986/87 Parliamentary session the Government will make a second attempt to confer on the prosecution the right to ask for a review by the Court of Appeal of any sentence which the Attorney General considers raises a question of public importance. Once again, the proposal is a compromise, falling short of giving the prosecution a straightforward right to appeal against the sentence on the ground that it is too short in the same way as the defence can appeal against the sentence on the ground that it is too long. The provision contained in the Criminal Justice Bill differs from the previous attempt in that the Court of Appeal will not carry out a post-mortem on the individual case, expressing an opinion on what the correct sentence should have been, but will confine itself to making a statement or reaffirmation of the principles that should be followed by sentencers in similar cases in the future. Leave to appeal will be required. As before, the individual offender whose case triggered the reference would not be at risk of having his sentence increased. The reasoning for this protection is that, if one offender whose case had caused a stir in the press were to be liable to have his sentence increased on appeal whereas another, whose offence and sentence were very similar but who had escaped public notice or criticism, and was as a result more favourably placed, then even-handed justice would suffer.

Despite the proviso that the Attorney General would limit his intervention to cases which he considers raise questions of public importance, it is hard to avoid the conclusion that the test of public importance will be the volume and intensity of public

---

[20] *Parl. Debates*, HL, 459 (5th ser.), cols. 386–406, 24 Jan. 1985.

protest. This is likely to prove capricious and unpredictable in operation, with the prosecution being drawn into sentencing decisions, seeking by advocacy to influence the sentence of the court, and introducing non-judicial yardsticks of leniency or severity resting upon some unknown and unknowable process of interpreting public feeling. No wonder that a former Law Officer speaking from the Government benches should ask the Home Secretary to think again.[21] The fallacy lies in trying to link the sentence of the court, and the process for review, to a perception of public opinion. This can only be a shaky and unsatisfactory basis for public policy. If a change is to be made it would be far better, in logic as well as in securing the declared aim of preserving public confidence in the penal system, to broaden the right of appeal against sentence into a two-way process, exercisable by the prosecution as well as by the defence. In Australia, four states—New South Wales, Victoria, Queensland, and Tasmania—allow appeals against sentence by the prosecution; so does Canada, where the practice is well-developed. No doubt the Crown Prosecution Service would make only sparing use of the facility to appeal where a sentence was markedly out of line with normal sentencing practice, but it would be a right extending to all cases, regardless of whether or not they had excited press or public attention. True, the convicted offender might have his sentence increased on appeal, but if in the view of the Court of Appeal the original sentence was too low, it is hard to see how there could be objection to a mechanism for upward adjustment in the interests of consistency, in precisely the same way as there is for reduction when the sentence is too high.

V

Belying an outward appearance of compactness and resistance to change, all the institutions and policies described in this book are inherently fluid. Like nuclei, they come together, divide and re-form in shapes that are sometimes familiar, but sometimes markedly different. Each of the institutions—Government,

---

[21] Sir Ian Percival, QC (Solicitor General 1979–83), in the Second Reading Debate on the Criminal Justice Bill. *Parl. Debates*, HC, 106 (6th ser.), col. 483, 27 Nov. 1986.

Parliament, and the courts—and their off-shoots are subject to
the flow of public opinion which, while it should never be
granted absolute supremacy (as demonstrated in the matter of
the over-lenient sentences), effectively moulds the policies and
shapes the systems that the politicians maintain. These opinions
may or may not be well informed, but they are the inspiration of
political action or conscious inaction. Penal philosophies have
been largely overtaken in recent years by practical policy
considerations. Gradually, there has been an ebb away from
the moral consensus shared by a small but influential group of
people of all political persuasions who took an enlightened
interest in questions of penal reform.

For half a century or more, judges and journalists, politicians
and civil servants—their consciences kept primed by the
venerable Howard League for Penal Reform and two newer
campaigners, the National Association for the Care and Resettlement of Offenders (NACRO) and the Prison Reform Trust—
have talked in much the same language when determining what
the objectives should be in drawing up just and effective penal
policies. The tone of the debate was distinctive, liberal and
paternalistic, a credit to the ideals of the penal reformers and to
the willingness of politicians and officials to listen and make
adjustments to the working of the penal system accordingly. The
abolition first of penal servitude and then of corporal and capital
punishment, the introduction of probation, parole, and community service orders[22]—all bear witness to the beliefs of the
informed and dedicated élite that concerned itself with penal
questions.

More recently, however, we have seen the previous orthodoxy,
and the idealism that prompted it, eroded by the steady,
seemingly irreversible, growth in the levels of reported crime.
Penal policy is now disposed to take account of the interests of
the victim as well as of the offender, while concepts of retribution
and deterrence attract much of the emphasis formerly accorded
to rehabilitation and reform. Nowhere is this better demonstrated than in attitudes towards treatment. Indeed, the word
itself has become almost a touchstone of the change that has

[22] Few penal innovations owe more to a single voice and personality than
community service orders owe to Baroness Wootton of Abinger. The origins are
described in Barbara Wootton, *Crime and Penal Policy*, George Allen & Unwin,
London, 1978, pp. 121–33.

occurred in public sentiment. From the end of the nineteenth century public policy had declared that convicted prisoners sentenced to imprisonment should be subjected to a regime of work, training, or education designed to 'turn them out of prison better men and women, both physically and morally, than when they came in'.[23] Over the years that followed, it became apparent that the ideal of rehabilitation was seldom realized, perhaps because expectations were pitched too high; whereas the reference to treatment carried overtones of sickness and cure. Yet crime is not a sickness, although some who perpetrate it are undoubtedly sick, and there are certainly few, if any, cures to be found in imprisonment.

It is hard to pinpoint what has caused the near abandonment of the belief in treatment in recent years. But in Britain as in the United States and elsewhere there has been a marked shift in attitudes, of the enlightened probably more than of the wider public, away from the aim of rehabilitating the offender towards the measure of retribution due to society as a consequence of the offence. That this is too blunt and indiscriminating an approach to the greater protection of the public will be argued in later chapters. While the remorseless increase in the levels of crime is undoubtedly the main cause, part of the explanation may lie in recidivism—that is, repeated offending by the same person over and over again, apparently unaffected by previous prison sentences or non-custodial alternatives—as well as by the distressing resort to violence. Research reports have also been pessimistic about the relative effectiveness of various forms of treatment assessed in terms of re-offending. Only too evidently, there is no panacea by which recidivists can be turned into law-abiding citizens by virtue of a few months or years in prison. So the secure containment of the persistent or high-risk offender has become an end in itself. At least, no further offences against the public can be committed during the time spent in custody.

In Chapter 10 I shall have something to say about the tensions between popular and informed opinion in shaping penal policy. For now, it is enough to remark that, as a result of the changing public mood and temper, most references to treatment are

[23] This wording was used by the Gladstone Committee in 1895. *Report from the Departmental Committee on Prisons* (C. 7702), Parliamentary Papers, V, 1895. Quoted in Nigel Walker, *Crime and Punishment In Britain*, University Press, Edinburgh, 1973, pp. 133–4.

virtually synonymous with the management or handling of convicted and sentenced prisoners. Although the move away from training and rehabilitation has combined with overcrowding and disturbances to alter the climate in the prisons, it should be said that opportunities for work, vocational training, and education still exist in the training prisons, and should be expanded. Even in a system that looks for its justification more towards retribution than reform, it is better to keep prisoners busy than idle. Why such a high proportion of offenders are sent to gaol, how they pass their time inside, and the case for organizational change are among the issues discussed in Chapter 7. With many thousands of prisoners now being released on licence each year, parole has become a significant ingredient in the penal system. Both the operation of the scheme and the justifications for it are considered in Chapter 8.

VI

As an assertion, I believe that few people would quarrel with the proposition that public concern about crime, and above all about violent crime, is widespread and deep-rooted. It cuts across income and class distinctions as well as party affiliations. There may be room for argument over the tenor of this body of public opinion, its origins, and its natural properties, as there certainly is over the way it is interpreted by politicians and in the press. But as to its existence, there can be little doubt. Popular reaction is kept stimulated by the detailed reporting of sensational and notorious crimes, often of murder, and sometimes of multiple murder. Behind these recurring peaks of public indignation lies a deeper and more constant feeling that the level of violent crime is intolerable, and that it is the duty of the responsible authorities to do everything in their power to counter the rise in crime, and to be seen to do so. As one of the most emotive currents of opinion in the modern body politic, it is a demand that politicians ignore at their peril.[24]

[24] On the day of his appointment as Home Secretary in Sept. 1985, Douglas Hurd unhesitatingly identified the campaign against crime and respect for the law as being the most important of his new responsibilities (*The Times*, 3 Sept. 1985). Law and order issues featured strongly in the Queen's Speech in Nov. 1985, while a Criminal Justice Bill was the centrepiece of the Government's legislative programme for 1986/7.

Like other popular beliefs, this current of opinion does not rest entirely on statistics but on more profound instincts. Unfortunately, no comfort is to be found in the criminal statistics, dangerous quagmire though they are to penetrate. The official statistics, published annually by the Home Office, are voluminous and can be interpreted in many different ways to support many different conclusions. The simplest, and most thought-provoking, formulation is that crime known to the police has increased sevenfold since the war. Less than half a million offences in the late 1940s had increased to a million by the mid-1960s, and had reached 2 million in the mid-1970s. The 1985 figure was in excess of $3\frac{1}{2}$ million. It needs to be repeated that offences recorded by the police are not a precise index of crime trends. As has already been shown, they do not reflect the full extent of crime, disregarding, as they do, a large number of offences that go unreported to, or unrecorded by, the police. Even recorded crime figures vary over time in ways that may not reflect actual offending patterns. Part of the rise in recorded crime, for instance, may be due to larger and more efficient police forces, the spread of telephones, changing patterns of insurance, altered recording practices, or a greater willingness on the part of the public to report crime to the police. Despite these factors, the conclusion that actual offending has increased, and increased sharply, is beyond all challenge. In part, this may be a response to enlarged opportunities for crime: more cars, more consumer durables of relatively high value, self-service in shops, and a more mobile population.

A large majority of offences are against property rather than the person, and many are of a comparatively petty nature. Fewer than 5% of all recorded crimes involve violence, but the crimes that do result in personal injury: the assaults, the woundings, and above all the robberies have escalated at rates comparable with (or, in the case of robberies, higher than) property offences recorded by the police. Killings also have increased, although mercifully less steeply. In the twenty years immediately after the war, offences of homicide initially recorded by the police fluctuated between 250 and 400 a year.[25] Since the mid-1960s, they

---

[25] Homicide in this context includes all offences of murder, manslaughter, and infanticide, but excludes causing death by dangerous or reckless driving (unless these are first classed as homicides), which was first distinguished from other manslaughters in 1956.

have risen steadily, now averaging some 600 per year.[26] To this total has to be added those who lose their lives on the roads as a result of reckless driving by others. In 1985 there were 290 court proceedings for causing death by reckless driving. Where convictions result, the casualties can only be regarded as victims of crime.

These sombre figures reinforce and illuminate public attitudes towards crime. Moreover, the scale of actual offending is greatly exceeded by the fear of potential offending. Fear of crime is rife, much of it out of all proportion to the real risks, but none the less emerging unmistakably from the British Crime Survey and other evidence as an unwelcome feature of contemporary society which cannot be overlooked. Especially in urban areas, there are now large numbers of people, many of them elderly and a majority of them women, who are living restricted and unhappy lives because they are afraid to go out in the evening, or even to leave home at all, for fear of being attacked or burgled.

Some consolation can be found in the fact that fear of crime and the chances of becoming a victim of crime are not related. Fear is concentrated among the elderly, the infirm, and the vulnerable, whereas the actual risks of personal victimization are highest for young men living alone in urban environments whose way of life involves such risk-enhancing activities as going out in the evening, particularly at weekends, using public transport at night, and heavy drinking.[27] The BCS and other surveys show that the elderly do not face the greatest risk from personal or household crime; in fact, the reverse is true—it is the young who are most at risk.[28] Valid though it is, however, this statistic may

[26] The average of 600 a year is the figure of crimes initially recorded as homicide; those that remain so classified after further investigation is more like 500. This is around 10 or 11 per million population a year, compared with 7 per million population in the late 1960s. The numbers convicted in each of the principal categories of homicide between 1981 and 1985 are shown in Table 6 in Chapter 5.

[27] Part of the difference in victimization rates between the young and the elderly may be explained by the fears of the latter. If they went about as fearlessly at night as their juniors, then it is likely that they would more often become victims.

[28] M. R. Gottfredson, *Victims of Crime: The Dimensions of Risk* (Home Office Research Study no. 81), HMSO, London, 1984, p. 30. See also J. Shapland, J. Willmore, and P. Duff, *Victims in the Criminal Justice System*, Gower Press, Farnborough, Hants, 1985.

do little to reassure the elderly and nervous who are apprehensive that the risk to them, albeit less than to other more robust groups in society, is serious and increasing. Particular caution is necessary if the growing emphasis on crime prevention is not to have the unintended effect of exacerbating the fear of crime among those who are already the most fearful.

One of the main purposes of *Responses to Crime*, as the title implies, is to relate the shifting but powerful currents of public opinion to the practical workings of the penal system—that is, to how offenders, once detected and convicted, are dealt with. Even before reaching that stage, when the damage has been done, opinion (local as much as national) may be activated to prevent the offender committing the offences that lead to appearances in court and what may follow. One of the most promising of all recent responses has grown out of initiatives taken in local communities to prevent or reduce the commission of crime in their neighbourhood. The potential for crime prevention and reduction, as yet only dimly perceived by the general public, is the subject of Chapter 9.

All these questions, which underlie the formulation of public policy, should be approached with an awareness that what is firmly lodged at the front of the public mind is that the level of crime is unacceptably high, that there is a paramount need for effective measures to curb the crime rate to be pursued with vigour and determination, and that the plight of the victim is more deserving of sympathetic attention than the needs of the criminal. The spontaneous growth of services for victims, and the enlarged role being sought for victims in the criminal justice process, as described in the next chapter, are proof of this concern. There are also subsidiary concerns about jury trials, the gross overcrowding in prisons, and the length of time spent on remand, but they are less dominant in the public consciousness. None of these propositions is as simple as it looks, but all have validity and deserve to be taken seriously. They are the elemental forces from which policies for dealing with crime have to be forged, and any proposals that fail to take them into account are unlikely to make much progress in the prevailing climate of opinion.

If we regard public opinion in the mass as those opinions held by private individuals which governments find it prudent to heed (which is still the best definition of public opinion),[29] we see in attitudes towards crime, and fear of crime, a conspicuous instance of a distinct body of opinion which bears directly on the policies of decision-makers. As Lord Scarman[30] has remarked, 'The truth of the matter is that the penal system, and the policies which inspire it, have got to remain within the toleration of public opinion.'[31]

[29] This classic definition was used by V. O. Key in *Public Opinion and American Democracy*, Alfred Knopf, New York, 1961, p. 14.

[30] Lord Scarman retired as a Lord of Appeal in Ordinary in 1986. As a High Court judge he had been chairman of the Law Commission (1965–73), and was a Lord Justice of Appeal between 1973 and 1977.

[31] BBC Radio 3, 17 Dec. 1984. Lord Scarman's comments were made when summing up a discussion following a broadcast lecture on 'Penal Policy and Public Opinion' by the author of this book. A shortened version of the lecture was published in the *Listener* (20 Dec. 1984).

2

The Victims

Nothing could be more even-handed than the statue of the Lady of Justice which presides serenely above the Central Criminal Court in the City of London. The notion of duality is inherent in the symbolism: the outstretched arms, the scales of justice, the sword. The message conveyed is that the sovereign state will weigh up the evidence and judge each offender according to the requirements of justice before imposing the punishment his crime has brought upon him. But nowhere in this awesome tableau is there room for the intrusion of a third party, the unfortunate victim. The physical arrangements inside the courts reflect the allegory without. There are places for judge and jury, for the accused, and for counsel, solicitors, police, and court officials. The professionals are at home in a familiar working environment.

For the man or woman who has suffered the wrong, extending from loss or damage to property to dreadful physical or emotional injury, there is nowhere to sit or stand or wait. Unless their presence is required as witnesses, no one pays any attention to victims attending a trial. Their reception varies from indifference at best to discourtesy at worst. This is not intended as a criticism of courts like the Old Bailey, where standards of administration are generally high, but as a truism applied to the criminal courts in general. The impersonal setting of court buildings is forbidding, and it is hardly surprising that so many victims experience disenchantment as a result of coming to court for the first time, or being involved at other stages in the sequence of events leading to the prosecution of an offender. Some end up vowing never again to offer their co-operation.

The justification usually offered for the lowly role accorded to the victim of crime is that his personal interests have been translated into a public cause of action, which takes into consideration not only the victim's wishes but those of the larger society.[1] On this view, the injury, loss, or damage sustained is a wrong not merely against the individual victim, but against the community as a whole.[2] Thus it is the community, through the agency of the state and its instruments for enforcing justice, which must be responsible for the application of sanctions, replacing the individual sufferer, vengeful or apathetic as he may be. This doctrine, while constitutionally impeccable, is of no great antiquity, nor does it represent a fixed and static relationship.

Before the middle of the nineteenth century, the victim was to all intents and purposes his own prosecutor, although prosecutions could be initiated by anyone. Then came the new local police forces. At first, police constables had no greater powers than the ordinary citizen, but the need for a power of arrest was soon recognized to further the aims of protecting the public and detecting crime. This in turn led to an expectation that the police would initiate and conduct prosecutions on behalf of the citizen,[3] the right of private prosecution being retained in reserve. Over the latter part of the century, the role of the injured party was steadily diminished. As a class, however, victims (potential as well as actual) benefited from a more orderly society. Reinforcing the evolutionary changes was a more radical concept: that the private citizens implicated on both sides of a criminal act—offenders and victims alike—had to take their place in a wider scheme of things, and that the organization of a civilized social life required the construction of a system of criminal justice with an identity and dignity apart from the conflicting interests of those who had the misfortune to fall under its mantle.[4]

[1] See A. S. Goldstein, 'The Victim and Prosecutorial Discretion: The Federal Victim and Witness Protection Act of 1982', *Law and Contemporary Problems*, 47 (1984), 245.

[2] The respective duties and entitlements of victims, offenders, and the state, and the issues of principle to which they give rise, are explored by A. Ashworth in 'Punishment and Compensation: Victims, Offenders, and the State', *Oxford Journal of Legal Studies*, 6 (1986), 86–122.

[3] See A. Sanders, 'The Prosecution Process', in D. Moxon (ed.), *Managing Criminal Justice* (Home Office Research and Planning Unit), HMSO, London, 1985, pp. 66–7.

[4] The establishment of police forces, the conditions in the prisons, and reforms in the courts are described by Sir Llewellyn Woodward in *The Age of Reform 1815–1870* (2nd edn), Clarendon Press, Oxford, 1979, pp. 465–73.

What happened next has been graphically described by a shrewd American commentator, Professor Abraham Goldstein. Although the role of the prosecutor has been historically different in the United States, the underlying principles and the course of development have been similar on both sides of the Atlantic:

It is, of course, true that the movement from private to public prosecution represents an advanced stage in the evolution of criminal justice. But public prosecution remains inextricably linked to the earlier stage in theory and in practice. It is the victim's injury and outrage which is, after all, 'channeled' into the public prosecution. In simple societies, this link between victim and public prosecutor is clear. The public official is a surrogate whose function is to provide assurance to the victim, his family, and his friends that they need take no punitive action and that the state will do what is necessary to bring the offender to justice. As societies become more complex—as more criminal laws are enacted for people who do not know one another and who are no longer organized to gather for vengeance and blood feud—the surrogate function becomes less visible and is almost forgotten. The prosecutor becomes another regulatory agency, administering a program of public control of criminal conduct in which policies of deterrence, rehabilitation, and retribution are to be weighed in the balance. He becomes increasingly susceptible to the exigencies of his situation—to the pressures of defendants, case loads, too many criminal laws, and unduly harsh sentences. As such, the prosecutor becomes less and less a representative of the victim. In short, his original 'channeling' function is displaced by a broader and more public function. A major tenet of the victims' movement . . . is the belief that the victim has been too much removed from the prosecution of the crime against him, that such removal underlies the common failure of victims to report crime and to cooperate in its prosecution, and that the prosecutor must somehow be recalled to his original function.[5]

In Britain, we have experienced over the last few years a definite shift towards taking more account of the interests of the victim as the third party in what is best portrayed as a set of triangular relationships between the state, the offender, and the victim. The figure overleaf illustrates in graphic form how the parties stand in relation to each other.

[5] Goldstein, 'The Victim and Prosecutorial Discretion', pp. 245–6. Abraham S. Goldstein is Sterling Professor of Law at Yale and a former Dean of the Yale Law School. His 1982 Law Memorial Lectures at the University of Mississippi on 'Defining the Role of the Victim in Criminal Prosecution' were reprinted in the *Mississippi Law Journal*, 52 (1982), 515–61.

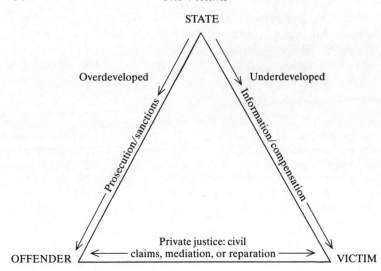

II

Financial compensation has always had a place in the settlement of disputes when the offender has had the means to compensate the victim for injury, loss, or damage caused by the commission of crime. Magistrates' courts may order a defendant to pay up to £2,000 in compensation upon conviction for a single offence, and they have jurisdiction to deal with most cases (save the most serious) of criminal damage, burglary, theft, and assault which come before the criminal courts. The Crown Court has unlimited powers. In cases of personal injury, direct compensation—paid by the offender to the victim—has been supplemented since 1964 by a non-statutory Criminal Injuries Compensation Scheme, one of the first in the world, which provides for *ex gratia* payments from public funds to the victims of violent crime.

In December 1983 the Government announced that it intended to introduce legislation to place the Criminal Injuries Compensation Scheme on a statutory basis. This would confer on eligible applicants a right to receive compensation and would provide specific parliamentary authority for the expenditure incurred. In total, some £220 million has been paid in compensation to more than 230,000 victims or dependants over the twenty-year life of the scheme, with the amounts increasing rapidly in recent years. The total for 1984/5, including the costs of administration, reached just under £40 million,[6] and this rose to nearly £46 million in 1985/6. It is a paradox that the very moment when this provision finally approaches the sanctity of the statute book, being included in the 1986/7 Criminal Justice Bill, should have coincided with an emerging scepticism in Whitehall as to whether resources will continue to be available to fund the scheme at its present levels. In the Second Reading debate on the Criminal Justice Bill, the Home Secretary described criminal injuries compensation as a new social service. Three-and-a-half times more was spent on it in real terms, he said, than in 1978/9. The number of people receiving compensation had risen far more rapidly than the increase in violent crime.[7] Consequently, the lower limit for claims on the scheme had been raised to £550 in order to concentrate additional resources (some £114 million over three years beginning in 1987/8) on those injuries that substantially affected people's well-being.

Apart from awards out of public funds, offenders are being ordered to pay compensation by the courts on a substantial scale. In 1984, 118,700 people convicted of offences in Magistrates' courts were required to pay compensation to victims, often in addition to other penalties, while the figure in the Crown Court was 6,200. The following year saw a decline to 114,200 in the Magistrates' courts, although the proportion of convicted offenders ordered to pay compensation was constant at about 15%. In the Crown Court there was a small increase in 1985 to 6,500.[8] If a court decides to impose a fine and compensation, and the offender has insufficient means to pay both, then it must give preference to the compensation order.[9] There is no way of knowing what effect insufficient means has on

[6] The number of cases resolved by the Criminal Injuries Compensation Board in the year ended March 1985 was 27,450 including 739 cases abandoned by applicants. In addition, 3,152 interim awards were made. Criminal Injuries Compensation Board, *Twenty-First Report* (Cmnd 9684), HMSO, London, 1985.

[7] *Parl. Debates*, HC, 106 (6th ser.), col. 471, 27 Nov. 1986.

[8] *Criminal Statistics for England and Wales 1985* (Cm. 10), HMSO, London, p. 171.

[9] Criminal Justice Act 1982, S. 67.

the size of compensation orders, or the extent to which these orders are enforced. Some research being carried out by the Home Office Research and Planning Unit should help to reveal more about current practice.

Impressive though they are, these statistics mask countless disappointments, lost opportunities, and unfulfilled expectations. When questioned, victims say they think that compensation should have a much more important role in sentencing than is currently the case. In the immediate aftermath of crime, however, thoughts of material loss are not always uppermost in victims' minds. It may be their predicament that predominates; a sympathetic and understanding response can be more important initially than talk of financial recompense. Police officers find that many people are still in a state of shock when an officer first meets the victim, and he or she may not want to, or be able to, hear about how to apply for compensation. A follow-up visit by a volunteer after more pressing matters have been resolved is preferable, provided the helper is adequately informed about the complexities of alternative remedies.

Crime, and the victims of crime, are infinitely varied, and in other cases the shock factor may be much less, and sometimes may be entirely subordinated to a desire to obtain compensation. It is interesting, although hardly surprising, that a survey in 1982 indicated that more than half of all the victims of violence who were interviewed definitely wanted to receive compensation of some sort, preferably from the offender by way of an order of the court, and as a right which they felt should be available to all victims as a consequence of what they had suffered.[10] The reluctance of some victims of violent crime to report incidents to the police means that compensation cannot reach all those who merit it. Even when it is reported, there is a disproportion that is hard to explain between convictions for criminal damage to property, which resulted in compensation orders being made in 64% of cases, and convictions for violence to the person, which

[10] See the summary of a report by J. M. Shapland, which was included as an appendix to the Minutes of Evidence which accompanied publication of the *First Report from the Home Affairs Committee of the House of Commons on Compensation and Support for Victims of Crime* (HC, 43), HMSO, London, 1984, pp. 20–2.

resulted in compensation orders being made in only 23% of indictable offences tried in the Magistrates' courts.[11]

Psychological and material factors are deeply entangled in the aftermath of a crime. Some victims, according to the National Association of Victims' Support Schemes (NAVSS), are insulted and compromised by any suggestion that money from an offender can put right the hurt and distress they have experienced.[12] Others feel they are receiving no more than what is due to them and are aggrieved at the small amount that may eventually be forthcoming. Even if the offender is caught and tried (and all the evidence points to the conclusion that only a minority are), the court will have to take his means into account. Consequently, the monetary sums available for compensation may be modest. Few offenders have the means to compensate victims adequately, and courts allow time to pay, perhaps at the rate of £1 or £2 per week, over several months. As a result, compensation orders can raise victims' expectations only to dash them, with small and irregular payments (sometimes falling short of the full amount awarded, for example when a persistent offender is reconvicted) serving to keep alive the memory of the crime for much longer than would otherwise have been the case. The same unpredictability does not apply to criminal injury compensation payments where they are granted, but there are qualifying limits and long delays. Despite the large number of claims received and awards made to victims and their dependants (nearly 20,000 in 1984/5),[13] the availability of the

---

[11] *Criminal Statistics for England and Wales 1985* (Cm. 10) p. 171. The fact that a higher proportion of violent offenders receive custodial sentences may have a bearing. See guidelines for the award of compensation orders summarized in *Stone's Justices' Manual*: 'It is rarely appropriate to add a compensation order to a custodial sentence as this may tempt the offender to commit further offences on his release.... Compensation orders ... [were introduced] ... as a convenient and rapid means of avoiding the expense of resort to civil litigation when the offender clearly has the means ... [to pay] ... In appropriate cases, courts will wish to draw the attention of victims of violence to the criminal injuries compensation scheme administered by the Criminal Injuries Compensation Board.' *Stone's Justices' Manual*, vol. 1 (J. Richman and A. T. Draycott, eds.), Butterworth and Co. and Shaw and Sons, London, 1986, pp. 443–4.

[12] *The Victim and Reparation*, a statement by the National Association of Victims Support Schemes, Apr. 1984.

[13] A total of 19,771 final monetary awards was made in 1984/5. A table published in the annual report of the Criminal Injuries Compensation Board showed the size of the awards. *Twenty-First Report*, p. 8.

scheme is still not as well-known as it ought to be, while arrears in investigating and processing claims means that an average claim is not settled for some nine months, a situation described by the chairman of the Criminal Injuries Compensation Board in evidence to a House of Commons select committee as 'appalling'.[14] It is to be hoped that the combination of additional financial resources and the new lower limit will mean that the Board will be able to keep its head above water in the processing of new applications, while at the same time making inroads on the backlog.

Thus, compensation, like much else relating to the unforseeable consequences of crime, is not so straightforward as it seems. The prudent citizen who has the means to do so ought to cover himself for as many risks as possible by insurance; but whatever the scale of financial resources available in the future, it is unrealistic to expect that any publicly funded scheme is going to be able to compensate every casualty of crime for all the distress, injury, or loss he may have suffered. Would that it could. The fact is that compensation has to be balanced with individual responsibility as well as measured against competing claims on public funds. Good causes are limitless; resources, by definition, are not. The numbers of orders made by the courts against offenders, where they have the means to pay compensation, are growing and are likely to continue to do so, but their relevance to victims is restricted to those cases where the offender is found, tried, and convicted. Compensation orders cannot touch the many offenders who are not caught at all.

The advance of the victim to the centre of the criminal justice stage, after remaining so long in the wings, should not obscure the fact that not all crimes involve an individual victim. Sometimes the loss or damage will be sustained by a corporate body, public or private, whose property is stolen or damaged. In large-scale frauds there can be many losers, and it is not always evident who has suffered loss and to what extent. Then there is a range of 'victimless' offences, such as the possession of controlled drugs or tax evasion, where an individual transgresses the law. Driving a motor vehicle without a licence or insurance

---

[14] *Compensation and Support for Victims of Crime*; see para. 30 of the main report, p. xii.

certificate does not involve a third party; nor, provided an accident is avoided, will there be any victim as a result of speeding, reckless driving, or driving after having consumed more than a prescribed amount of alcohol. Yet all are crimes and can in some instances attract heavy penalties. Incitement may be a crime even if the person incited does not in the event commit any crime; while sexual offences like soliciting in the street, indecent exposure, or obscenity are still crimes, irrespective of whether anyone is attracted or repelled, disgusted, upset, or depraved. Whether or not a prosecution follows in the case of a victimless crime (and here it is the quality of life of a wider public that is affected, rather than that of an individual) will depend on whether it comes to the notice of the police or other prosecuting authority, and what view is taken of the gravity of the offence. But every crime of violence to the person necessarily produces a victim in its wake, and it is the more widespread recognition of this social evil that, more than anything else, has led to the emergence of new attitudes towards victims.

III

Since the mid-1970s, in Britain and the United States, as elsewhere in North America and Western Europe, the minimal role of the victims of crime has come under increasing scrutiny. Although the developments have been running in parallel, the origins are different. In the United States the first impetus came from attempts to strengthen the prosecution process by encouraging witnesses to give evidence in court, aiming for a higher rate of convictions and what could be presented as a more efficient system of criminal justice. With the opportunities for political advantage coinciding with humanitarian instincts, it was not long before a new and potentially powerful body of opinion was created, supplemented by social and community pressures. Thus, concern over rapes, child abuse, wife battery, and violence inside and outside the home caused the women's movement and other voluntary agencies to join in campaigning for the provision of free assistance and support for the victims of crime. The climate of opinion was also influenced by victimization studies making more vivid the huge amount of unreported crime, including crime among ghetto-dwellers (as distinct from crime committed by blacks outside the ghetto).

In 1978 some existing voluntary programmes in Brooklyn and elsewhere in New York City (initially aimed at benefiting the prosecution) were taken over and expanded by the city when Mayor Koch inaugurated the Victim Services Agency. The original emphasis on the victim as witness has been maintained and broadened in the provision of services to help alleviate the psychological, financial, and practical problems that victims and witnesses face in their encounter with the bewildering criminal justice system. Counsellors familiarize complainants with court procedures, meet and look after them when attending court, and act as a channel for ensuring that they attend on the day and time they are required. This notification process, alerting police as well as civilian witnesses, has greatly reduced, if not completely eliminated, much unnecessary time and expense spent in wasted attendances on occasions when witnesses were not needed. In 1984 the Victim Services Agency calculated that police officers were spared about 114,000 unnecessary trips to court in the year at a saving to the New York City Police Department of more than $18 million in resources. By calling victims to court only when their appearance was required, the notification process helped civilian witnesses avoid more than 72,000 unnecessary appearances. Advance notification can also help to ascertain if financial restitution is sought, or if there are practical problems bearing on attendance in court such as absence from work, ill-health or immobility, child care or transport.

Once the victim or witness arrives at the court building, he or she is directed to a reception centre or a separate waiting area staffed by voluntary or paid counsellors, several of whom will be bilingual in the Spanish-speaking areas of the city. Reception centres are intended to be places to wait in comfort, to drink coffee, or make telephone calls, where young children can play or older ones can be supported if they are called to give evidence in court. Above all, they enable witnesses who have been victims of crime to be kept in the picture, and to be told what is going on and what is expected of them. American experience has shown that it is important for reception centres to be kept apart from the public areas, if only by partition screens, so that they are not accessible to defendants, or to the families, associates, or friends of the defendants. Intimidation of witnesses, overtly or more subtly, or even unexpected confrontations between defendant and victim, can constitute an acute problem during the lengthy periods of waiting that characterize most criminal trials. Secure facilities of this sort are highly desirable, both in terms of avoiding psychological shock or distress and, in the event of intimidation, reducing the possibility of interfering with the course of justice.

Additional services provided by the Victim Services Agency to the courts include the assessment and collection of financial restitution paid by the offender to the victim if the court so orders, and the preparation of victim impact statements which can be used in plea negotiations and sentencing decisions. Much of this may sound foreign to the practice in our own courts, but, if so, it is to our disadvantage. At the very least, there is a lot to learn, and to adopt, in ways of receiving and looking after victims and witnesses when attending court, as there is from an improved system of prior notification. Quite apart from the worthwhile contribution that such measures can make towards countering feelings of alienation experienced by victims caught up in the criminal process as a consequence of crime, more attentive attitudes towards victims and witnesses should encourage the reporting of crimes and a closer co-operation with the enforcement authorities. As to the cost, I shall later put forward a suggestion which could provide an additional source of funds.

Outside the courts, the work of the New York Agency is closer to our own victims' support schemes in local communities, although impressive in its range and inventiveness. In four large city hospitals, counselling and emotional support is provided for victims of rape, sexual assault, and domestic violence when the hospitals' own social service staffs are off-duty. Twenty-four hour telephone hotline services are available to victims of domestic violence and young people who have run away from home, and are well used by both of the groups for whom they are intended. In 1984 the Agency gave assistance and support to no less than 13,000 victims of domestic violence, most of them battered women, 6,600 of whom had first made contact by the hotline. Other programmes cover special aid for the families and dependants of homicide victims; support for elderly people who

have suffered injury at the hands of an unknown offender or
have been abused within their home by another family member
(it is significant that as many as one in every twenty-five
Americans over the age of sixty-five is abused in this way); and
specialist services for children who have been, or are at risk of
being, victims of crime.[15] Children invariably blame themselves
for the crimes inflicted on them, and if (as is commonly the case)
the assailant is known or related to the victim, the child feels
betrayed and disoriented as well. Parents feel guilty for not
protecting their child, and may not know how to react at the
moment of crisis or afterwards. And for most children who are
called to testify in court, the experience can be intimidating and
confusing.

IV

Simply to enumerate some of the services existing in New York,
one of the more violent cities known to man, is to illustrate the
extent of victims' needs and what can be done to meet them.
These same emotional and practical needs were present in
British society in the 1970s—not so highly coloured, perhaps, or
so articulately presented, but there all the same. From the start,
the emphasis in Britain has been on the plight of the victim as an
individual who has suffered from crime, rather than as a
potentially useful witness for the prosecution. In 1974 an experimental scheme was launched in Bristol. Previously the Chief
Constable had said that the police would co-operate with local
volunteers in a pilot project in one police division when he was
satisfied that responsible volunteers would be selected, and that
they would receive some instruction and training from professional sources. Early decisions were that there should be a co-
ordinator to act as a contact point with the police for referrals, as
well as to collect information about victims and allocate work to
volunteers. Since crime occurs 365 days of the year, and it was

---

[15] The US Attorney General's Task Force on Family Violence reported that
spouse abuse, incest, child molestation, battering of children, and abuse of
elderly victims, once thought of as isolated events occurring among only a small
element of the population, represented widespread problems that occur among
families in every social and economic class. *Attorney General's Task Force on
Family Violence, Final Report*, Washington, September 1984.

regarded as essential to offer help to victims quickly, the scheme itself would need to operate every day. As a crisis service, it would concentrate on immediate relief and assistance and would not attempt to provide long-term support. Where there was a need for continuing help over a longer period, the victim would be referred to the appropriate statutory or voluntary service.

Behind these functional decisions, which set the pattern for the hundreds of other local schemes that were to spring up throughout the country, was the belief that victim support should be local and voluntary, not just as a practical and money-saving way of meeting the very real needs of the casualties in their midst, but as an expression of community concern. The earliest enquiries had established that many of the victims of crime had not simply suffered loss or injury, but were hurt by the feeling that the local community did not want to know about their experiences. Neglected and sometimes isolated, the realization that no one seemed to have a responsibility towards them could heighten a victim's distress. Natural anger about the crime could turn into bitterness towards the authorities and the wider community in general.

Like many a pioneering effort, the first victims' support scheme in Bristol, despite help from NACRO and some local well-wishers, encountered initial setbacks and had to suspend its work after nine months owing to lack of funds. Fortunately, some detailed research had been carried out during the first six months which demonstrated the value of the project. Nearly 400 victims had been identified, mainly through referrals by the police, and visited by volunteers as soon as possible after the commission of the crime. In about half of these cases, serious practical and emotional problems existed which were capable of being eased by some form of outside assistance. After some timely publicity on a BBC television programme, interest in victims' support began to spread more widely, and within the next two years eight further schemes, all modelled on the Bristol prototype (by now happily revived), were set up in different parts of the country.

As the network grew, information and experiences were exchanged and common attitudes began to emerge. By 1977 representatives of some existing local schemes were meeting regularly in the knowledge that the number of potential schemes was increasing rapidly and that several of them were facing difficulty in getting off the ground. In the following year a steering committee was elected charged with establishing a national association. When planning the formation of an association, the founding members determined not only to offer a service to all bona fide schemes, but also to set minimum standards of good practice as a protection for the victims they sought to serve. Essential criteria were that the schemes should be competently managed; that they should earn the co-operation of the local police force; that volunteers should be carefully selected and suitably trained; and that appropriate records should be kept. It was implicit that all schemes should be non-profit-making, non-political, and non-sectarian. Representatives of the Association of Chief Officers of Police and the Chief Probation Officers Conference joined in, and finally, in August 1979, the National Association of Victims' Support Schemes was registered with the Charity Commission. Shortly after obtaining charitable status, the Home Office and DHSS made grants to augment income from trusts and private donors to enable the recruitment of a paid office staff. The initiative had been taken by fifteen local schemes out of approximately thirty which were then in existence. Over the next six years, the growth was to be momentous.

Although the work of supporting victims has remained essentially voluntary in character, the statutory agencies, notably the police and the probation service, have been ready enough to co-operate with the volunteer schemes in their locality. The police, usually first on the scene after a crime has been committed, have welcomed outside assistance to take on the task of consoling and supporting victims and their families. Whereas sympathy and companionship may be the first priority in many cases, there are practical matters to be attended to as well. Claims may need to be made and forms filled in and sent off to an insurance company or the Criminal Injuries Compensation Board; statements may be required by the police; contact made with relatives and friends; local authority housing departments or landlords notified of damage to property; builders found to carry out repairs; perhaps transport arranged and appointments made at a hospital or other centre for professional care or treatment.

Over a short period of years, the infectious growth in the network of victims' support schemes, now covering most of the country, has been an inspiring story, a milestone in the honourable history of voluntary social service. In its (sixth) annual report for 1985/6, the NAVSS was able to record that nearly 4,000 people were working as trained volunteers, backed up by a further 2,815 helpers serving on management committees or providing professional advice and liaison with other agencies. In 293 towns and districts in England, Wales, and Northern Ireland there were affiliated local schemes, with additional schemes in Scotland and the Republic of Ireland under the auspices of their own national associations. Between April 1985 and March 1986, 184,994 new victims and their families had been referred to support schemes, an increase of 47% over the previous year.[16] In that year, the Archbishop of Canterbury had indicated his personal support by becoming President of the Association and contributing a thoughtful foreword to its annual report.

It is all a far cry from the cavernous criminal courts in Brooklyn and the hotlines of the Victim Services Agency in New York and other American cities, but equally laudable in its aims and achievement. At a time when criminal justice has seen so many adverse trends, the way in which organized victims' support has expanded, bearing the unmistakable marks of a genuine grass-roots movement, stands out as a shining and largely spontaneous example of resourcefulness and magnanimity towards a hitherto neglected group of individuals in distress.[17] The human spirit is fortified by this contemporary example of how man is still capable of helping his fellow men, as he always has, and of doing so, moreover, without the intervention of the state until the mould has been set.

V

Mediation and reparation take the aim of righting wrongs a step further. This is uncertain ground, lying at the base of the triangle

[16] National Association of Victims' Support Schemes, *Sixth Annual Report 1985/86*, London, 1986, p. 16.

[17] In addition to the local victims' support schemes described in this chapter, there are a number of other statutory and voluntary services available to victims. For example, *The Times* reported on 27 Nov. 1985 that 45 rape crisis centres were in existence, with more opening. These provide an initial sympathetic contact by telephone, supported by individual counselling for victims of rape.

where the moral and personal obligations due from one citizen to another, to whom he has caused loss or injury, are to be found. These obligations will remain latent, however, unless activated by information received by the victim from the relevant agency of the state. Viewed in this way, information about a case is the precursor of interpersonal contact between victim and offender. It is regrettable that so little has been done on any systematic basis to keep victims better informed of the progress before the courts of cases in which they have a direct interest. Unless he makes enquiries, or is required as a witness, a victim or member of his family may not know if 'his' offender has been identified and charged, nor will he necessarily be notified of the outcome of the trial. A powerful example of the resentment and frustration that can result is given in the second of the two case histories that are recounted in Chapter 4. As in pre-trial notification of witnesses, there is ample scope for improvement in the flow of information to other interested parties, which modern technology now facilitates. Far more reports and records should be handled electronically in order to relieve the overloaded administration of the courts, and at the same time keep victims and witnesses in the picture.

The aim of mediation—to ease the path of the victim between the time of the crime and the moment, weeks or months later, when the offender is convicted and sentenced[18]—is not helped by the fact that, in most of the experimental mediation and reparation schemes, the process of mediation does not start until a guilty plea or judgment has been entered and social enquiry reports have been requested by the court. This is the practice in all of the Home Office funded experimental schemes that were set up in 1985 save one, that in Carlisle. There, the probation service is testing a pre-trial scheme which is aimed at diverting juveniles from the criminal justice system altogether. The police refer to the scheme selected offenders aged between fourteen and sixteen who have admitted to offences, usually of theft or damage to property. The probation service then tries to negotiate reparation with the victim, the offender perhaps making good the damage he has caused or offering some other form of service in

---

[18] The average waiting time for criminal cases to come to trial in London and the South-east was reported to be 11½ weeks for defendants in custody, and 17½ weeks for those on bail. *The Times*, 27 Aug. 1986.

recompense. If reparation can be agreed and satisfactorily accomplished, the probation service may recommend that the police should caution the offender rather than prosecute him. By resort to this procedure, the victim is helped and the young offender is prevented from taking the first fatal step on a penal ladder that only too often leads onwards and upwards from an initial appearance in the juvenile courts to more and more serious criminal offences and sentences in the years ahead.

In the case of adults, the mediator has to be on his guard against coming between the offender and the court. The functions of conviction and sentence are exclusively the prerogative of the courts. Most magistrates and judges in the Crown Court, I suspect, would subscribe to the view, already discussed, that, as a crime is an offence against the wider community, it is for the authorized instrument of the state, namely the court, to decide on the punishment. Yet few would adhere to this purist stance so far as to preclude a thorough enquiry by the court, after guilt has been established but before sentence, into the background and motives of the offender. The type of sentence and its impact are crucial factors affecting future offending, and it is in this context especially that mediation and reparation can make a constructive contribution to the purposes of criminal justice.

The role of the mediator is a most delicate one in terms of human relations, both with the offender and with the victim, in eliciting the information that will be placed before the court prior to sentence. It is important to realize that, far from being a soft option, for an offender to meet his victim, if the victim is willing to do so, is often much more difficult than for the offender to make an appearance in court. Appearing in court, very possibly not for the first time, can be little more than a formality. Even persistent offenders typically do not comprehend legal proceedings. They are told when to stand up and when to sit; for much of the time they do not feel personally engaged, shrugging off their involvement in the ritual of justice as no more than part of the game.

By contrast, there is no room for remoteness or feelings of otherworldliness in a direct and sometimes intense meeting with the victim. Here the offender must participate with all his faculties in an acutely sensitive interchange. Resolution and courage are required to meet the victim face to face, seeing just what it has meant to have property damaged or stolen, or how in the more extreme cases a victim's way of life or health may have been shattered by the crime. The impact on the offender can be profound, and probation officers say it is only in mediation proceedings that a certain sort of offender is brought to realize that another human being was harmed by his criminal act.

If a victim is prepared to meet 'his' offender, it is to be hoped that he has got over any feelings of overt bitterness or hostility, although there can be no certainty about this. For some, the anxiety and distress caused by the crime may be relieved by a personal confrontation. Victims may want to know what the offender is really like and if he may come back and harm them again. Why, in particular, did he single them out as a target? Does he have criminal associates still at large? Had the house been watched? Questions like these may not always be put into words, but they are invariably present. A meeting may resolve several unanswered questions, particularly since a high propor-tion of burglaries are in fact opportunist crimes—not planned in advance against any particular target, but with the house selected by chance simply because it looked empty and easy to break into. Reassurance on this score may help to console the victim and restore his peace of mind.

The psychological consequences of crime on victims are still only dimly comprehended. The word 'traumatic' is overused, but it does describe fairly enough the experiences and reactions of victims who become lonely and isolated—not daring to go out because they cannot come to terms with the realization that another human being, whom they may have known or who may be a stranger, has committed a crime against *them*. Not all of these social casualties are the most obviously vulnerable—the single, elderly people living alone: victims of all ages and backgrounds can suffer, with the mental effects surviving for an almost unendurably long time afterwards.

VI

Although it is early days yet for mediation, a case history will bring the process into relief. A probation officer visiting an elderly victim learned of the circumstances in which she had had her pension money stolen shortly before. The victim was a seventy-seven-year-old widow, deaf, partially sighted, and at times confused. Nevertheless, she was able to give a coherent account of the incident. What had happened was that, after she had been to collect her pension, she was walking home when a youth ran past her and grabbed her handbag. A second youth ran off with the assailant, but some passers-by saw what had happened and gave chase. Eventually the youths threw the bag down, where it was picked up and returned to its owner. The woman was very shaken by the incident, but fortunately she had not lost her balance or been knocked over. While she was not physically hurt, she was emotionally upset and agitated. Neighbours had come to her assistance and the victims' support team had been called in. At the first meeting she had asked about the two youths, wondering if they had done it before, and if they would do it again.

Later, after the police had arrested and charged two men, who turned out to be brothers in their early twenties, the offenders expressed remorse for what they had done, saying they could not understand what had made them do it. Subsequently they had experienced nightmares, and, in the case of one of them, vomiting. Although living in the locality, neither was previously known to the police or the probation service. Both pleaded guilty, and the probation service was asked for social enquiry reports. Since the men had offered to apologize to the victim, the case was referred to the mediation scheme probation officer who had already visited the elderly woman. The probation officer discussed the possibility with the victim, who replied she would like to meet them, not in her own home, but at a day centre she was attending and with some of the social services staff present to give her moral support. Accordingly, a meeting was arranged, and the two brothers were brought to the day centre by the probation officer.

After being introduced to the victim, both men apologized readily, asking after her health and her condition since the crime. One sat next to her and took her hand, saying later that she reminded him of his own grandmother. The elderly woman told them how she had been shocked and distressed, and that since the incident she had not felt able to go for her pension herself which a relation now collected for her. But she said she accepted their apology and was pleased to have the chance to meet them as it made her feel less frightened and more secure, especially as they were so evidently upset. The mediating probation officer reported that the meeting was an emotional one; all parties were very nervous beforehand and one of the men ended in tears. Both were visibly shaking, repeating over and over again, as they had from the time of their arrest, that they could not think what on earth possessed them to do what they had done. They told the reporting officer that they found the experience an ordeal but worth-while, and said they would like to do some voluntary work at the day centre as reparation, if their services could be used. The woman victim expressed herself well satisfied with the meeting, telling the staff that it had restored some of her trust in young people. The professional staff concluded that her level of anxiety had been beneficially reduced as a result of the encounter.

The report for the court by the probation service recommended that the possibility of a community service order should be considered in view of the brothers' remorse, their apology, and their willingness to make amends for their behaviour. Moreover, both were first offenders who had pleaded guilty. They had been assessed as suitable for community service, and it was proposed that, with the agreement of the social services department of the local authority, the period of community service should be performed at the day centre where they had met the victim. It was suggested that this penalty would reinforce the lessons that the defendants had already so painfully learned and would enable them to make amends for their actions.

When it came to deciding on the punishment, however, the Magistrates' court declined to accept the recommendation, sentencing both men to six months' imprisonment, although one of the defendants had the custodial element of the sentence suspended after an appeal to the Crown Court.

The implications of this narrative are worth thinking about. In the event, the Court took a different view from the probation officer. But that is nothing unusual. The responsibility of every criminal court is to balance what it sees as the interest of the public in terms of retribution and example with the rehabilitation of the offender. The scales of justice are not comfortable to hold.

The probation service, on the other hand, will naturally want to do all it can to lead an offender away from criminality and towards a more law-abiding life in the future. That, too, is in the public interest. What the episode does clearly demonstrate is that, in those limited areas where mediation and reparation are practised at present, there is no question of a private bargain being struck between victim and offender in terms of reparation and apology setting aside the interests of the wider community as interpreted by the courts.

So did mediation fail in this instance? Surely not. The experience of meeting the woman in person and seeing her and other elderly and infirm people at a day centre can only have left a profound impression on the two men. The act of atonement was virtuous in itself, and genuine compassion seems to have extended from and to both parties. Who can doubt that one elderly woman slept better at nights and walked out with more confidence by day as a result of her meeting with the shadowy and threatening figures who had snatched her handbag?

VII

When the first support scheme was taking shape in Bristol, an unexpected problem occurred in identifying the victims. Many of those who had suffered from crime were shaken and resentful. Other than the police, if the crime had been reported, or sometimes relatives or neighbours, the plight of the victim—indeed, even his existence—might not be known. As time went on, with more and more victims' support schemes being established, the availability of the service became better known, with a growing volume of referrals resulting from the police and the public. Word of mouth also played a part, especially in urban areas.

While the idea of reparation as an enforceable sanction, as distinct from acts of reparation arising out of mediation, is still embryonic, the practical difficulty in finding suitable tasks has recurred. Some advocates, while not going so far as to support the idea of reparation orders made by the courts, none the less see in reparation a promising alternative to court proceedings, especially where they end in custodial sentences. The overcrowding in the prisons and the delays in the courts are postulated elsewhere as justification enough for trying to divert as many offenders as possible away from the penal system. Few Home Secretaries indulge in radical notions, but in March 1984, when announcing that the Home Office would be willing to fund a small number of victim–offender reparation schemes on an experimental basis, Leon Brittan recognized 'a feeling that justice would be better done if some way could have been found to require the offender to make some positive recompense to the victim or the community as a whole'.[19] The next day, the Parliamentary All-Party Penal Affairs Group published a report on victims calling for the response to crime to be recast to place 'the interests of the victim at the forefront of our consideration'. The first recommendation was that reparation (including compensation) should become a central feature of sentencing policy.[20] The approach is an interesting one which has attracted considerable discussion, but it is unlikely to make much contribution to easing the congestion in the courts or the prisons. Nor does it take account of the objection that a criminal offence is committed not simply against the individual victim, but against the community at large. This is more than an entrenched and hidebound reaction: it has contemporary relevance to the situation of the victim, his more prominent role in the process of criminal justice, and his relationship with the offender. Throughout, it is important to avoid spreading the impression that the victim is simply being 'used' to further the interests of the offender.

Most of those working actively for victims are cautious about putting too much reliance on mediation and reparation. So are some of the probation officers who are engaged in experimental schemes in the field. In April 1984 the NAVSS reported that, despite the favourable publicity that had accompanied developments presenting mediation and reparation as a course of action in the victims' best interests, many members of victims' support schemes had difficulty in seeing what benefit there could be for the victim. The NAVSS statement recognized the potential benefit to the criminal justice system in diverting offenders from

[19] Speech to the Holborn Law Society, 14 Mar. 1984. Quoted by T. F. Marshall in *Reparation, Conciliation and Mediation* (Research and Planning Unit Paper no. 27), Home Office, London, 1984, p. iii.

[20] Parliamentary All-Party Penal Affairs Group, 'A New Deal for Victims', mimeo, 1984.

imprisonment, and also pointed out that direct contact with victims could play a constructive part in the rehabilitation of offenders and consequently in the reduction of crime in the future. The idea that the offender was in conflict with the state rather than with another individual citizen, the Association added perceptively, made it too easy for him to deny responsibility. Offenders frequently attempt to justify their acts by claiming that they have not harmed anyone by theft or criminal damage (assuming that insurance will replace stolen items with something better), whereas those with whom they are dealing after sentence, such as probation or prison officers, will seldom know about the circumstances of the crime or the impact on the victim and so are unable to counter these attitudes. Personal contact with the victim can bring home to an offender, in a way that court proceedings never can, just what the effects of his behaviour have been on the life and security of another individual.

Where the actual victim does not want to meet the offender, or become involved in any way, it is sometimes possible for someone else to act as a go-between in negotiations leading to an apology. This may be followed by a period in which the offender gives assistance to others who have suffered from crime, or to those in need for other reasons. Also, there appears to be a sufficient number of victims who are prepared to visit prisons or youth custody centres in groups so as to discuss with convicted and sentenced prisoners what it feels like to be on the receiving end of criminal activities. Victims do not meet their 'own' offenders in these encounters, which to some of those taking part may have a therapeutic value, allowing victims to express their feelings and to convert a negative experience into something more positive. Occasionally a victim may be motivated to take an interest in offenders generally and in the aim of rehabilitation. For offenders, also, meetings with those who have suffered from crime may provide an opportunity to talk and explain their conduct, to be questioned about it, and perhaps so to understand better the impact of what they have done and the anguish caused to innocent victims. No one with first-hand experience of prisons and prisoners would want to pitch too high a claim for the value of meetings of this sort, even when they are attended by members of the staff of the institution or the police. Nor, so far as I am aware, are they organized on any widespread scale. But at least they are an indication of a willingness on both sides to think and talk about what causes people to commit criminal offences, and to appreciate more fully what the consequences can be to the victims as well as to the offender who is caught and punished.

In an attempt to discover the extent and range of mediation and reparation projects, the Home Office co-operated with the NAVSS on a survey conducted in the spring and summer of 1984. With the approach to classification and definition being described as 'pragmatic', the survey identified a total of forty-two systematically organized and continuing reparation/mediation schemes, supplemented by a further sixty-eight at planning stage. These schemes varied considerably and were listed in detail in a published report.[21] Only in six out of a total of forty-two areas in England were projects neither in existence nor planned. At the time of the survey, twenty-five schemes were operating in seventeen mainly urban areas of the country as far apart as Exeter and Northumbria.[22] There was no standard pattern, as had been the case with victims' support schemes, although certain common features, such as reference from the police or the courts, stood out. Some schemes had been launched by local residents, religious bodies (e.g. the Society of Friends in Edgware), or existing voluntary organizations (e.g. Save the Children Fund in Lambeth). A larger number originated from initiatives by the social services department of the local authority, the probation service, or the police. In most instances there was a close liaison with the local victims' support group.

The projects themselves were divided into those where the offenders made reparation to their own victims, described as 'direct' reparation, and those where the reparation was made to other victims of crime, described as 'victim assistance' schemes. Community service, whether ordered by the court or otherwise, was excluded, as it did not normally involve any encounter between offenders and victims. On the basis of this survey, and

[21] T. F. Marshall, *Reparation, Conciliation and Mediation.* A revised version by T. F. Marshall and M. E. Walpole was published in 1985 under the title *Bringing People Together: Mediation and Reparation Projects in Great Britain* (Research and Planning Unit Paper no. 33), Home Office, London, 1985.
[22] Ibid., p. 9.

indeed from practical observation generally, it is clear that direct reparation, although intellectually the most satisfying, is in practice rare. Apologies may be offered or negotiated, motives varying from a genuine desire for forgiveness and reform to a more prosaic hope that the court may be sufficiently impressed by an apology and offer of reparation to ameliorate the sentence. But individual victims can be reluctant to come forward. Many, probably a majority, may feel uncertain about such a difficult and nerve-racking encounter, even if arranged by a mediator who is present. Some will simply not want to know, being suspicious of pressure to accede to a deeply unsettling experience merely on the outside chance of rehabilitating an offender. Moreover, practical tasks of direct reparation are hard to find. Even when property has been damaged, it may have had to be repaired more quickly. A broken window cannot wait for some weeks before the offender and victim agree on reparation. Nor is the offender necessarily the most appropriately skilled person, or the most reliable, to carry out repairs or make good damage. Then there are questions of insurance, and how claims may be affected, and the attitudes of neighbours or others living in the house. The evaluation of the four experimental schemes being funded by the Home Office—those operated by the statutory services (in Coventry, Leeds, and Carlisle), and one managed by a voluntary organization (the Crypt Association in Wolverhampton)—will enable a more informed assessment to be made of all these difficulties.

It is presumably due to a combination of practical and psychological factors that most of the schemes surveyed in 1984 offering victims direct reparation by their 'own' offenders were characterized by a very low take-up.[23] The wider, if less personal, alternative of victim assistance seems to be more promising, although even here there are practical difficulties.

In September 1985, David Mellor, a Home Office minister, observed that reparation, which had originally been thought of as more suitable for relatively mild offences of theft or criminal damage, might be applicable in more serious cases as well. He added that reparation schemes were being introduced in response to public demand, and that he believed they filled a gap

---

[23] Ibid., p. 37.

in the criminal justice system.[24] But with the change in Home Secretary from Leon Brittan to Douglas Hurd in the autumn of 1985, combined with the scepticism of many of those working in the field, the signs were that the tide was turning. A discussion document issued by the Home Office described the four experiments and invited comments on reparation, either as an alternative to prosecution or as an order of the court.[25] While there is still considerable interest in enhancing the role of mediation and reparation in the criminal justice system,[26] there are powerful arguments of principle, as well as practical and procedural difficulties, against making reparation the subject of court orders. Instead of blurring distinctions and provoking unnecessary political opposition, it is wiser to rest upon the concept of the triangle, where the sentence of the court is the reflection of the offence committed against society, with mediation and reparation forming part of the relationship of private justice linking the offender and his victim.

VIII

Politicians as a class, on both sides of the Atlantic and in all parties, have not been slow to see the potential of the victims' movement. From deep inside local communities, in themselves valuable quarries for votes, there has been a spontaneous response by a growing number of well-meaning private individuals to the acute needs, not hitherto separately perceived, of those who have suffered personal injury or loss or damage from crime. Because of their misfortune, victims rightly attract a great deal of sympathy, and credit is well-earned and generously conferred on those who are ready to help them in their distress. Where such fertile ground exists, the seeds of political exploitation are always likely to flower. In the United States, the

[24] Speech at The Crypt Association, Wolverhampton, on 16 Sept. 1985.
[25] *Reparation: A Discussion Document* was published by the Home Office in 1986.
[26] *Reparation and Mediation*, a bibliography prepared by S. Prashker and N. Brearley, was published by the Forum for Initiatives in Reparation and Mediation in 1986. Marshall and Walpole, *Bringing People Together*, also included an annotated bibliography.

stentorian words of the introduction set the tone of a report to the President:

When you established the President's Task Force on Victims of Crime on April 23, 1982 you led the nation into a new era in the treatment of victims of crime. Never before has any President recognized the plight of those forgotten by the criminal justice system—the innocent victims of crime...

Citizens from all over the nation told us again and again how heartened they were that this Administration has taken up the challenge, ignored by others in the past, of stopping the mistreatment and neglect of the innocent by those who take liberty for license and by the system of justice itself.[27]

The desire to gain political capital is natural enough, but the measures proposed, and the advantages claimed, need to be scrutinized with a cool head and a sharp eye. This applies similarly to special interest groups, whether campaigning for women's rights, the restoration of capital punishment, the greater accountability of the police, or any one of a host of other libertarian or moralistic causes. Each of these may detect advantages to be obtained in aligning themselves with the sympathetic figures of the victims of crime. All such groups are entitled to their views, and to disseminate them by whatever means are open to them. But they should not seek to exploit victims in order to attract support for the policies they advocate. Those services that exist primarily for the benefit of victims, such as victims' support schemes, would do well to think carefully before joining forces with any such groups.

The early creeds of victim support—to be non-political and non-sectarian—have endured unchanged over the first decade and should be preserved. The third limb of the trilogy—to be non-profit-making—has seldom if ever been a problem for local schemes struggling to raise enough money to survive. Central government funds amounting to £126,000 in 1985/6 were made available to support the cost of the NAVSS national head-quarters, representing about 60% of the total. The annual report of the NAVSS for the same year estimated that grants to local schemes from public funds totalled £1,041,372, the principal donors being county councils and other local authorities, the

---

[27] *President's Task Force on Victims of Crime, Final Report*, Washington, December 1982, p. ii.

Urban Programme, and the Manpower Services Commission.
This was an increase of around one-third compared with the
previous year, but the number of schemes, and hence the volume
of work undertaken, had also increased steeply.[28] In the previous
year the time and effort spent in obtaining local grants, coupled
with their unpredictability and short duration, had led the
NAVSS to seek a more secure base by way of central government
funding to meet the cost of a co-ordinator and the administrative
expenses of each affiliated local scheme.[29] The proposal won
endorsement from the House of Commons Home Affairs
Committee in its first report in the session 1984/5.[30] In reply,
the Government laid emphasis on the independence of local
schemes and the importance of the spontaneity and enthusiasm
of the voluntary effort on which they depended.[31]

When giving evidence to the Select Committee, Leon Brittan
had argued that the voluntary character of local schemes was
intrinsic to the way in which victims' support had developed. It
had not originated from the Government saying 'this ought to be
done; let us find the people to do it', but had been a genuine and
spontaneous initiative from inside local communities.[32] While
well disposed towards the NAVSS and its affiliated schemes, and
greatly valuing the work done, he had not been able to accept it
as either right in principle or realistic financially for central
government to assume responsibility for funding all local co-
ordinators and administrative costs. At a time when money was
short, he said, large sums were being directed towards victims
who had sustained personal injury by way of criminal injuries
compensation payments, amounting in total to £39.5 million in
1984/5, while the voracious demands of the police, the courts,
and the prisons had to be contained within an overall policy of
not increasing public expenditure in real terms. As a token of

[28] NAVSS, *Sixth Annual Report 1985/86,* pp. 26–7.
[29] Some local schemes had paid co-ordinators, either part-time or full-time,
although the amounts paid varied considerably. Where the MSC funded a co-
ordinator it was normally for only one year. Of all local schemes in 1984/5, 55%
were co-ordinated by unpaid volunteers (see NAVSS, *Sixth Annual Report 1984/
85,* p. 25).
[30] *Compensation and Support for Victims of Crime.*
[31] *Compensation and Support for Victims of Crime: the Government Reply to
the First Report from the Home Affairs Committee, Session 1984–85 HC 43*
(Cmnd. 9457), HMSO, London, pp. 67–89.
[32] *Compensation and Support for Victims of Crime,* pp. 116–27.

good will, however, the Home Secretary agreed to consider the possibility of some limited assistance to local schemes. This undertaking was subsequently honoured by the provision of £136,000 in 1986/7 to establish a contingency fund to assist local schemes that were experiencing financial difficulties. In its annual report the NAVSS, which was to be responsible for determining applications, commented that the money would be a lifeline for many schemes. It added, however, that a proportion would undoubtedly be used to pay for essential items previously hidden, rather than for spearheading new developments in victims' support.[33]

What happens in practice has been the subject of a thorough investigation commissioned by the Home Office from the Centre for Criminological Research at Oxford.[34] The study assessed in some detail what volunteers actually do for victims, what victims think of the service offered or provided, and whether it measurably assisted in their recovery. Samples of victims were questioned, and the conclusion was that a great majority had welcomed visits by volunteers, with a substantial minority believing that the support or assistance received had significantly affected their recovery. The success of letters and telephone calls, either as expressions of caring or as ways of persuading seriously affected victims to ask for a follow-up visit, was more limited, although some victims were reached by these means who would otherwise have received no attention at all.[35] Evaluated in terms of utilization of resources, schemes operated more effectively in making the most of their limited amounts of money than in their deployment of non-financial resources, notably available volunteers.[36] The effectiveness of schemes, not surprisingly, tended to wax and wane with the energy of dedicated individuals, above all the co-ordinators. The broad conclusion of the study was that there was a genuine and widespread need for support schemes and that their work was important to victims, especially in the most demanding area of the long-term support of those who have suffered from serious crimes. For these victims in

[33] NAVSS, *Sixth Annual Report 1985/86*, p. 28.
[34] E. M. W. Maguire and C. L. Corbett, 'The Effects of Crime and the Work of Victims' Support Schemes', Final Report to the Home Office, April 1986. To be published by Gower Press, Aldershot, Hants, in 1987.
[35] Ibid., p. 277.
[36] Ibid., pp. 278–80.

particular, a more professional organization was required. In its present state of development, the case for government funding, at least for large urban schemes, was seen as being very strong.[37]

In the light of this report, and other evidence, it seemed plain that the funding of voluntary victims' support was haphazard and unlikely to keep up with the expansion that will inevitably result from the growing practice of automatic rather than selective reference of cases by the police, as well as the need for more specialized support services.[38] While there was a danger of evolving into one more paid social service, bureaucratized and professionalized in a way that would be probable if victims' support were to rely entirely on full-time paid staff, none the less, the short-term and piecemeal nature of public funding was beginning to hold back further progress, besides taking the edge off the initial enthusiasm of volunteers. There was also a growing credibility gap between the warmth and volume of ministerial praise and the paucity of material support. An awareness of these factors enabled Douglas Hurd to persuade his ministerial colleagues to make available £3 million a year in new money for three years from 1987 onwards. The extra funds should allow the co-ordinators to be paid in many of the more active local schemes. The allocation of the additional funds, announced in October 1986, was to be agreed between the Home Office and the NAVSS.

The decision marks a clear and welcome departure from the previous policy of making contributions to headquarters' expenses, backed up by periodic hand-outs to schemes in financial difficulty. Central government has now indicated a readiness to maintain victims' support as a regular and specific commitment, rather than on an incidental or exceptional basis. These developments supersede a different scheme, put forward by the Howard League for Penal Reform in the report of a committee chaired by a High Court judge, Sir Derek Hodgson. Their report envisaged the establishment of a fund to aid victims from the proceeds of crime when confiscated by the court in

---

[37] Maguire and Corbett, op. cit., pp. 293b–c.

[38] The Metropolitan Police informed the House of Commons Home Affairs Committee that since February 1984 it had been the policy of the force to refer automatically all victims of certain categories of crime. *Compensation and Support for Victims of Crime*, p. 101.

victimless offences, supplemented by fines paid to the court and compensation orders.[39] The proposal foundered on the rocks of Whitehall on two counts: first, because there was no correlation between the amount that might be paid in and demands for payment out, so that the scheme could be a net but unpredictable drain on resources; and, second, because of the general Treasury opposition to hypothecation.

There is, however, another precedent in the United States that cries out for notice and consideration. In many states public funds are now provided to assist the victims of crime from general revenues and fines imposed by the courts, as well as from penalties assessed on convicted offenders. In 1984 penalty assessment schemes were operating in twenty-two states, sixteen of which also made provision for victim assistance from fines. In that year the National Organization for Victim Assistance (NOVA) reported that:

Funding legislation is often difficult to pass since states face increasingly critical economic decisions. However, the trend has been to fund victim services through penalty assessments or fines on all convicted offenders. This funding source is increasingly used for victim services and victim compensation. This means that such victim funding is generated primarily by imposing burdens on those convicted of crimes, and not from state government funds.[40]

The cost of fixed penalty assessments to the convicted offender was in the range of $12.50–$30.00 in 1984, with a higher rate applying to counts of felony than misdemeanour. By mid-1986, however, in one state at least—Colorado—legislation had been passed increasing from $75 to $100 the cost levied on felony actions resulting in convictions, to be paid into the victim compensation fund.[41]

Penalty assessments may be waived by the court, and often are in the poorer districts where an indigency procedure allows relief to those without the means to pay. This proviso is sensible, since otherwise further sanctions, maybe even imprisonment, would

---

[39] *Profits of Crime and their Recovery* (Cambridge Studies in Criminology, Vol. LII), Heinemann, London, 1984.

[40] *Victim Rights and Services: A Legislative Directory, 1984*, NOVA, Washington DC, p. 4.

[41] National Association of Attorneys General, *Crime Victims' Report*, June 1986, p. 9.

result, and the time of court administrators would be wasted in pursuing in vain the non-payment of small sums. In contrast to the criminal courts in Brooklyn, where penalty assessments are waived for a large majority of those convicted of criminal offences, in the more prosperous up-state areas of New York a high proportion of all convictions (including numerous motoring offences) attract penalty assessments paid to the court. For motorists in particular, a down payment of a relatively small amount represents only a fraction of the costs of running their car, perhaps less than is spent weekly on garaging or having it cleaned. Few problems are encountered in collection, the amount being paid to a court official immediately at the conclusion of each case.

It is important to grasp that penalty assessments are not part of the sentence, nor are the proceeds paid directly to the victim (if there is one). In this respect they differ from the Hodgson proposal. They are a compulsory surcharge, authorized by the state legislature, imposed on all convicted offenders on their passage through the criminal justice system, provided that, in the opinion of the court, they have the ability to pay. In some states the proceeds are used to offset the cost of court administration, which may include counselling and reception arrangements for victims/witnesses as well as prior notification services. In New York City, the Victim Services Agency is partially funded in this way through funds transferred from the state to the city. Elsewhere, penalty assessments are paid into separate funds to benefit or compensate the victims of crime. Penalty assessment schemes, or various forms of fine/surcharge arrangements, are now commonplace in the United States. Unlike the much more narrowly conceived orders made under our Costs in Criminal Cases Act 1973, whereby a defendant may be required to pay the prosecution costs, surcharges represent a deliberate method of redistributing funds from one class of person (the offender) to another (the victim) via a public agency. At a time when all revenues resulting from general taxation are under extreme pressure, it is a device that has much to commend it.

Why could it not be done here? Opposition, on ideological as well as practical grounds, comes from an unexpected quarter: those who stand closest to victims and offenders and have most to gain. Some of the leading personalities in victims' support tend to shy away from anything that might polarize victims and offenders. Apprehension is expressed that an additional financial penalty would be regarded as an extra punishment, not only by the offenders themselves, but also by many probation officers and offender-based charities. It is claimed that surcharges would also work against the principle of interpersonal mediation, where an offender is encouraged to face up to the responsibility he has for his own behaviour, rather than simply pay for the harm he has done. Then there is the consideration that victims' support schemes have become wary of anything that can be divisive between the interests represented on their local committees, such as the police, the probation service, and voluntary organizations. It has not been easy to bring the different parties together, and the key to success has been in focusing on the main issue on which all are agreed: namely, the provision of support for the victims of crime. What all would like to see is greater provision made for victims' services from central and local government funds.

Although there is a natural worry about becoming too. dependent upon a precarious and unspecified amount arising from surcharges or penalty assessments from year to year, the present arrangements, improved as they are, do not hold out the prospect of being able to meet all the demands of victims' support. Many of those working for the cause share a yearning that victims' support should be regarded as an essential public service like the police, fire, and ambulance, which ought to be adequately funded by the state on a permanent basis as an integral part of its obligation towards its citizens. Forcefully argued as it is, this is special pleading, albeit for a good cause, whereas officials in the Treasury and elsewhere in Whitehall have developed thick skins and hard hearts. In the end, it is the politicians who decide on what resources should be allocated to what purposes, but they too have to reconcile a national policy to restrict public expenditure in the interests of containing inflation with the particular claims of a multitude of conflicting special interests. For this reason, if no other, any gift horse that appears on the scene should be very carefully examined before being sent away.

IX

Many jurisdictions in the United States, federal as well as state, have now enacted provisions for the better treatment of victims of crime. While it is hard to particularize victims' needs, and harder still to translate them into rights, there is an assumption that needs extend beyond support and assistance. Two propositions in particular have crystallized: first, that victims are entitled to be respected and heard; and second, that they need to be informed about how their loss or injury can be mitigated, about where they can find immediate or long-term relief from the problems caused by the criminal act, and about the progress of the case if an arrest has been made. 'Being heard' may vary from an obligation placed on the prosecutor to consult the victim of a serious crime, as in the Federal Victim and Witness Protection Act 1982, to enabling the victim to address the court before sentence. The federal statute provides for consultation at four stages: dismissal of the charge or prosecution; release of the accused pending judicial proceedings; plea negotiations; and pre-trial diversion.[42] Victim impact statements, by means of which the court is acquainted with the victim's assessment of the crime, are now fairly common and generally welcomed. There is less enthusiasm for going further and permitting victims to ask the court to take their wishes into account in deciding on the sentence. To go down this road would be to introduce another entirely subjective element into sentencing which would be impossible to maintain or defend.

Variations in practice are a healthy sign that the public response to the situation of victims is still evolving. In Britain, marked progress has been made with the rapid multiplication and increasing professionalism of support schemes; interesting and worthwhile experimental work is being done in the fields of mediation and reparation; while compensation, whether from

---

[42] See Goldstein, 'The Victim and Prosecutorial Discretion', pp. 229–32. The Victim and Witness Protection Act 1982 directs the Attorney General of the United States to develop guidelines for the Department of Justice in order to obtain the views of the victim of a serious crime, or his family, at specified stages of the criminal process. It also requires that a victim impact statement be included in the pre-sentence report prepared by the probation officer for the sentencing judge. These provisions apply only to criminal cases brought in the federal courts, although there is comparable legislation in many states.

public funds via the Criminal Injuries Compensation Scheme or direct from offender to victim, is now operating on a massive, although still far from comprehensive, scale. We have, in our reserved British way, some considerable distance still to go in reaching out to the victim in the criminal process. It is now high time to devote ourselves singlemindedly to strengthening contact with victims, explaining what is happening and why, and listening with respect to what they have to say. Improved reception arrangements in the courts should be a priority. None of this need be expensive; there is scope for volunteers, and the costs of a national scheme could easily be met by penalty assessments. Indeed, the probability is that there would be a comfortable surplus for other purposes connected with court administration and services.

Only by these means can the bitterness, resentment, anger, and frustration that are so often the consequences of crime be assuaged. These are emotions that impoverish the character of those who have already suffered enough. They are also deep among the root causes of alienation from the criminal justice system. Doing more for victims is not simply the right thing to do on humanitarian grounds: it can go a long way towards relieving penal policy of some of the burdens which it carries.

3

Living off Crime

A single day at the Old Bailey will encompass the extraordinary diversity in motivation and circumstance in the most serious criminal offences. In Court Number One a murder trial is in progress in which the accused are charged with killing a police officer pursuing a Scotland Yard investigation into an organized robbery of three tons of gold bullion worth £26 million from a security warehouse at London Airport: an all-time record haul. In an adjoining court a group of young men, Londoners of West Indian descent, are also facing murder charges, this time as a consequence of a stabbing incident at an all-night party that got out of hand. The defendants were caught up in an affray in which knives and broken bottles were used by opposing factions; much alcohol had been consumed, and tempers flared after racial taunts had been thrown about. After a painstaking summing up in which the judge carefully explains the law of murder, and the degree of intent required, the jury finds sufficient grounds of provocation or self-defence to return verdicts of not guilty of murder, but guilty of manslaughter or affray. Prison sentences of four years or three years result for the leading characters, with lesser penalties for those who played a minor role. Some of the defendants are acquitted by the jury.

The contrast between this excitable brawl and the case unfolding in Court Number One is striking. There, it is alleged by the Crown, an undercover police officer carrying out surveillance in the course of an extensive police operation had been stabbed to death when discovered in the grounds of the home of one of the defendants. Both defendants are suspected of being deeply involved in organized crime. They have the appearance of intelligent and capable men, well-dressed and alert. Their defence by a leading member of the criminal bar is skilful and ultimately successful. The jury, each member of which has been under police protection twenty-four hours a day since the trial began in order to prevent any attempts at intimidation or inducement, finds them not guilty of murder. The judge has already ruled that the Crown has the burden of disproving defence counsel's submission that the men were acting lawfully in their own self-defence having been attacked by an unidentified and aggressive intruder. The police officer is dead and there are no other witnesses. In the event, neither defendant is released at the conclusion of the trial, both remaining in custody to face further charges arising out of the disposal of the stolen gold.[1]

A few months later they are back again in the dock in the Central Criminal Court. Once again the two men, by now accompanied by other co-accused, plead not guilty. For eleven weeks another jury is under police protection, and after giving their verdict jury members are released from jury service for twenty years. The verdict, reached after deliberations lasting more than thirty-six hours, is that both of the defendants are guilty of conspiring to handle the bullion. The principal defendant, Kenneth Noye, who is shown to have masterminded a huge laundering operation to recycle melted-down gold bars on to the legitimate gold market, is sentenced to fourteen years' imprisonment and fined £500,000. Civil proceedings are launched without delay for the recovery of the gold and the proceeds from its sale. Assets estimated to be worth at least £3 million were frozen, while simultaneously the Customs and Excise takes action for the recovery of up to £1 million in unpaid VAT. The Inland Revenue follows up with a writ for £949,000 in respect of tax, national insurance, and interest due between 1978 and 1984.[2] The second defendant, described as Noye's vigorous right-hand man, received a total of nine years' imprisonment. The trial discloses some surprisingly accommodating banking practices, one suburban High Street branch having accepted £1 million in £50 bank notes paid into a newly opened bank account by Noye, using a false name, over a period of three days.

The transnational character of organized crime was well

[1] For a police account of the bullion robbery and the death of the police officer during the subsequent investigation, see Gilbert Kelland, *Crime in London*, Bodley Head, London, 1986, pp. 208–11.

[2] The *Observer*, 27 July 1986.

illustrated by this case. Almost three years after the bullion robbery, a special unit set up by Scotland Yard to track down the laundering process utilized to change bullion into cash, and cash into investments or property, was reported to have uncovered a massive international network of shell companies and accounts handling millions of pounds of American as well as British criminal profits. The links with organized crime in the United States embraced drug traffickers and dishonest lawyers, with enquiries extending from Florida, Boston, Chicago, and New York to banking centres in Latin America, Europe, and the offshore financial havens of the Caribbean.[3] Large-scale organized crime is invariably ruthless and determined, although fortunately it is comparatively rare for deliberate murder to feature in the planning. This may be due less to any particular regard for the sanctity of human life than to its irrelevance as a means to securing the primary objective: money. The desire for profit, so deep-seated in human nature, and sometimes linked to the pursuit of power or authority, is the mainspring of organized crime. Yet since callousness and a disposition to use whatever degree of violence may seem to be necessary in attaining the objective, injuries leading to incidental loss of life can and do result.

The term 'organized crime' is in general currency and is useful enough as a description. But it means different things to different people and deserves some thought and analysis, both as a phenomenon to be studied and understood and as a collective wrong to be countered in the most effective way. Before singling out some of the characteristics of organized crime, I should say that in what follows I have excluded politically motivated terrorist crimes. It is not always obvious where the line should be drawn between 'ordinary' crime, that is, non-politically motivated offences, and those that form part of a planned campaign aimed towards a political end. Most people recognize a terrorist incident when they see one—indeed, the propagandists for the cause may proclaim it. It is a remarkable vindication of the versatility and inherent strength of the courts and penal institutions that, in mainland Britain at least, it has been possible to try those accused of political offences in the same courts and according to

---

[3] *The Times*, 10 Nov. 1986.

the same procedures as other offenders. Nor has it been the practice for convicted terrorists sentenced to imprisonment to be held in separate gaols. In Northern Ireland emergency powers have been taken to enable a judge to sit without a jury, deciding questions of fact and law in those criminal cases where the process of justice is likely to be distorted by intimidation or reprisals on witnesses or jurors.[4] The so-called 'Diplock' courts, and the conditions and regimes at the Maze and other prisons[5], have been accepted reluctantly as necessary by governments of both parties over the last decade. Nor has the practice been very different in the Republic of Ireland, where three judges sitting without a jury try scheduled cases of a political character. The recourse to these measures has been criticized, particularly by those sympathetic to the interests of the minority in Northern Ireland, but they have withstood the challenge of numerous appeals to various international tribunals and bodies. Nevertheless, it is right that periodic scrutiny should be given to the continuing need for such departures from normal practice.

II

Organized crime is hard to pin down. In its essentials, it amounts to the activities of any group of individuals who are organized to profit from the community by illegal means on a continuing basis. This formulation contains three elements: that the activity should be organized, that it should be illegal, and that it should be continuing. The degree of organization and structure will vary, as will the scale. A useful distinction now used by the police, but originating from a criminologist, is between 'craft' crimes and 'project' crimes:

The *craft* organization, typical of people performing skilled but small-scale thefts and confidence tricks, is a small, fairly permanent team,

[4] Non-jury courts for the trial of terrorist offences in Northern Ireland were authorized by statute in the Northern Ireland (Emergency Provisions) Act 1973, later replaced by the Northern Ireland (Emergency Provisions) Act 1978). Known colloquially as 'Diplock' courts after the Lord of Appeal who had recommended their introduction, they were the subject of a notice of derogation by the United Kingdom from the provisions of the European Convention of Human Rights. This was withdrawn in 1984.

[5] See *Report of an Inquiry by HM Chief Inspector of Prisons into the security arrangements at HM Prison, Maze* (HC 203), HMSO, London, 1984.

usually of two or three men, each of whom has a specific role to play in the routinised thefts in which the team specialises. It is a team of equals and the profits are shared equally at the end of each day. The *project* organization typical of burglars, robbers, smugglers or fraudsmen, engaged in large scale crimes involving complicated techniques and advance planning is an *ad hoc* team of specialists, mustered, sometimes by an entrepreneur, for the specific job in hand. Profits are shared on a basis worked out beforehand, though some participants may work for an agreed flat fee.[6]

The higher reaches of organized crime may embody a pyramid of authority ranging from the precarious ascendancy of one or more criminal gang bosses—sometimes partners in crime, but sometimes rivals when they fall out or their interests diverge—to the more permanent hierarchies found for example in the control of vice. Continuity and a sense of shared purpose distinguish organized crime from individual offences which are committed in response to events or on impulse when opportunities offer. The composition, even the existence, of a grouping may not be constant; individuals with special skills for a particular job coalesce and then disband. A majority of them, however, will combine more than once over a period of time to plan and carry out criminal activity. These collective projects are defined and held together by a shared purpose—each member with his place, each performing a particular role. The law's response to the bond of shared purpose is the offence of conspiracy.

American usage embraces all of these characteristics, but goes further in identifying linkages between groups of differing ethnic or geographic origin, and a greater readiness to resort to overt fear and violence in the pursuit of illegal goals. The amazing extent and ramifications of the Mafia families emerge from time to time as a New York or Chicago godfather lies dying in the street where he has been shot, more often by gangland rivals than by the forces of law and order. In the United States the spread of organized crime is recognized as stretching far beyond those traditional, but by no means defunct, Italian/American conspiracies, the Mafia and Cosa Nostra. In the words of a presidential commission 'Organized crime is not merely a few

---

[6] M. McIntosh, *The Organization of Crime*, MacMillan, London, 1975, pp. 28–9.

preying upon a few. In a very real sense it is dedicated to subverting not only American institutions, but the very decency and integrity that are the most cherished attributes of a free society'.[7]

At a more mundane level, despite herculean efforts to clean up the cities, literally as well as metaphorically, waste disposal in New York is reputed to cost up to half as much again owing to the intervention of racketeers; while gambling, the construction industry, the docks, funeral homes, car dealers, wine and spirits distribution, and vending franchises are among the areas of legitimate entrepreneurial business which are regarded as particularly prone to penetration by organized crime.[8] A study by the Rand Corporation, commissioned by the National Institute of Justice, has substantiated the encroachment of organized crime upon businesses ranging from oil distribution to cheese manufacturing to garbage collection. The problem, according to the Rand survey, seems generic to 'industries characterized by small firms, low technology, services provided at the customer's location rather than the vendor's, and a strong union'.[9]

It is too comfortable to think that it cannot happen here. The more lurid manifestations of Mafia rivalries may appear to us little more than bizarre caricatures, stemming from alien cultural traditions which are to be found in parts of Europe as well as in the New World. But deviant instincts are no different, while the international currency of drugs combining with faster and less rigorous methods of handling money—electronic transfers, offshore banks, the buying and selling of currency, and the vast extension of credit—all help to facilitate the laundering of ill-gotten gains on a world-wide scale. Translating the proceeds of crime into businesses, cash, houses, land, or moveable property which can be enjoyed in safety is crucial to those who engage in organized crime. Without the means of doing so, many (although not all) of the incentives would disappear.

[7] *The President's Commission on Law Enforcement and the Administration of Justice: Task Force report on Organized Crime*, US Government Printing Office, Washington DC, 1967; quoted by D. C. Smith, 'Organized Crime and Entrepreneurship', *International Journal of Criminology and Penology*, 6 (1978), 175.

[8] The *Observer*, 22 Dec. 1985, carried a report on the activities of some of the leading Mafia families in the United States following the shooting in New York of Paul Castellano.

[9] The Rand Corporation, *Criminal Justice Research at Rand*, Oct. 1985, p. 27.

Criminal intelligence, as well as journalistic and other sources, suggest that large-scale organized crime in Britain is for the most part centred in London, but with thriving underworlds in other cities including Glasgow, Liverpool, and Manchester. Those living on the proceeds of crime, either directly or when they have been suitably laundered, concentrate their energies on a limited number of criminal activities which hold out the prospect of a high rate of return: robbery, theft of valuable loads when in transit, counterfeiting, fraud, drug trafficking, and arms dealing. Of these, drugs and fraud are currently likely to produce the greatest returns for the resources invested and the risks involved. But the picture changes. At various times in the past, high profits have been achieved from extortion, as well as from pornography and gaming. Recently, as a result of greatly improved security precautions, especially by the tobacco and wine and spirits industries, there has been a decline in thefts of high-value loads.[10] Fraud, however, seems to be a growing and boundless source of criminal income. The appalling social consequences of the trade in dangerous drugs is only too apparent, triggering a secondary wave of lesser crimes carried out by drug-takers to finance their addiction. Arms dealing provides links to politically motivated international terrorist groups, as well as with others set on destabilization for their own political ends.

Most of the larger-scale organizers do not confine themselves to just one type of crime. It is more profitable to indulge in several at the same time. Thus, the proceeds of one venture can be used to finance another. As in legitimate business, diversification pays dividends. Nor is this the only comparison that can be made between organized crime and legitimate business. The objective of both is profit. In organizational terms, vertical integration is the byword. Market research, planning, operations, investment, financial control, and legal advice all have a place in identifying a market and exploiting it to the full. Much ingenuity is displayed by organized criminals, and legitimate businesses

---

[10] Yearly comparisons are provided by the Tobacco Advisory Council, Security Liaison Office. Gross losses adjusted to 1969 prices show a decline, although the 1985 total was the second worst since the adoption of improved security measures. See the Council's Security Liaison Office's 'Report for 1985' (mimeo).

may be set up as fronts engaging in legal trading in parallel with laundering the proceeds of crime or pursuing other objectives an illegal way. In the nature of things, the anatomy of organized crime is seldom exposed, although occasionally as in the case of Charles Richardson, a south London businessman who was convicted in 1967 and sentenced to twenty-five years' imprisonment, a trial in the courts illuminates the cruelty and extent of a notorious criminal conspiracy.[11] The Kray twins too were sentenced, this time to life imprisonment with a recommendation by the judge that they should each serve a minimum of thirty years, after a sensational trial. Apart from the killings that led to their life sentences, it transpired that much of the twins' support for charity and sporting events, as well the operation of their socially fashionable clubs and restaurants in the West End and elsewhere, depended on protection rackets extorting money from clubs, public houses, and betting shops. There was no mercy for those who failed to pay up.[12]

The economic impact of organized crime is impossible to calculate, but Britain is a prosperous society, with large sums of money within reach of the unscrupulous. Detected cheque frauds alone are estimated at between £50 and £60 million in 1985. This statistic covers losses incurred by UK banks participating in the cheque guarantee card scheme and by the major credit card companies, but excludes cheques used to defraud private individuals rather than the banks. The police believe that only in a small proportion of cheque frauds is the person carrying them out arrested and charged. Currency counterfeiting is now conducted on such a scale that the Bank of England discourages disclosure of the annual amounts of sterling counterfeits that are recovered, lest confidence in the currency be undermined. But an occasional glimpse of the magnitude of organized forgery comes to light, as in December 1985, when a single police operation in central London resulted in the arrest of three men and the recovery of nearly £1 million in face value of counterfeit notes

[11] R. Parker, *Rough Justice*, Fontana, London, 1981, contains an account of Charles Richardson's rise and fall.

[12] N. Lewis, *Britain's Gangland*, W. H. Allen, London, 1969, describes the ramifications of the Krays' criminality in detail at pp. 195–304. Also see J. Pearson, *The Profession of Violence: The Rise and Fall of the Kray Twins.* Weidenfeld & Nicolson, London, 1972.

together with equipment for forgery. In addition to the sterling counterfeits, there is a world-wide business in counterfeiting US dollars carried out in this country, as elsewhere, to say nothing of other counterfeits which are neither identified nor recovered. We must hope that few incidents of alleged deception rival the dimensions of a venture which led to the appearance of four businessmen in court in 1986 on committal proceedings, charged with conspiracy to obtain £16.6 million by deception by claiming to be able to obtain anti-tank missiles for Iran.

Drug trafficking in all its labyrinthine forms is an extremely profitable business, and it is estimated that in 1985 something like £500 million was required to fund the consumption of dangerous drugs.[13] The outlay and risks incurred by the principal organizers of the distribution networks are low, and the profit element will have represented a large percentage of the total turnover. In 1985 Customs seizures at the seventy points of entry to the United Kingdom accounted for a further £107 million at street prices, and the Chief Investigating Officer of the Customs and Excise emphasized that this represented only part of the traffickers' attempts to import drugs illegally.[14] International co-operation had prevented still more controlled drugs from reaching Britain, but inevitably a substantial and unknown quantity had eluded the controls. Once these drugs have entered the country, an informed guess is that about one-half of the total volume of drug trafficking takes place in London, with rather more than half of the organized distribution groups being based there. A fuller account of the consequences to the drug-taker at the end of the chain follows in Chapter 6, together with a review of the main strategies for tackling drug misuse.

Fraud can be even more profitable than drugs, although until recently it has enjoyed a lower public profile and may command greater degrees of skill and ingenuity. Fraud investigations are time-consuming and lengthy, making heavy demands on police manpower. Outside London, police resources to combat fraud are slim; in a medium-sized force in 1985 it was unusual for more than a dozen officers to be working full-time on the

---

[13] This estimate was given by Assistant Commissioner John Dellow of the Metropolitan Police in September 1986. *The Times*, 23 Sept. 1986.

[14] Ibid., 8 Jan. 1986.

detection of fraud.[15] Too often, police have had to rely on preliminary investigations carried out by the insurance companies, or by the banks or accountants, with the results passed to the police or Director of Public Prosecutions for them to decide whether or not there was a basis for prosecution. In London the resources are greater, but so too is the volume of offending, with an international perspective to the largest frauds. For example, one investigation during 1985/6 was into the London end of a Malaysian/Hong Kong fraud believed to involve the equivalent of £600 million. Extortion of money from businesses, sometimes backed up by kidnapping or threats of violence or product contamination,[16] is another instance of an ageless form of criminal activity that has adapted its methods to the ways of modern living.

Apart from the police, other enforcement agencies are engaged in countering large-scale fraud. The 1985 Report of the Customs and Excise, for example, revealed the vast scale of VAT frauds connected with dealings in gold:

Altogether 39 persons were convicted in respect of evasion totalling £14,410,000. They were sentenced to terms of imprisonment ranging from 3 months to 6 years, were fined a total of £318,000 and were made the subject of Criminal Bankruptcy Orders totalling £14,178,000. However, there is considerable doubt whether much of this £14 million or so will be realised. Although there is no direct evidence, all the indications are that the major portion of the fraudulent profits are in the safe haven of inaccessible foreign bank accounts.

A number of earlier investigations had not come to trial by the end of the year. A further 35 persons have been committed on charges alleging the evasion of a total of £33,834,000. In one of these cases the committal proceedings lasted more than 12 months; possibly the most protracted series of committal proceedings ever to take place in English courts. During the year a further 13 persons were charged with gold frauds involving the

[15] Appendix G to the *Fraud Trials Committee Report*, HMSO, London, 1986, showed the strength of fraud squads in the police forces of England and Wales at pp. 231–2. The total number of officers concerned solely with the investigation of fraud in 1985 amounted to 588.

[16] A new criminal offence to strengthen the law in dealing with contamination or threats to contaminate consumer goods, with penalties up to five years' imprisonment, was enacted in the Public Order Act 1986.

alleged evasion of £5,910,000 VAT. During the course of these investigations gold to the value of £145,000 was seized.[17]

III

Organized crime on this scale—the deliberate, systematic, and continuing criminal activitites of a consistent group of people—is the work of little more than a few hundred individuals, not usually related to one another, mostly living in or near London, although with occasional forays away from home when projects are carried out in the provinces or abroad. Many more make a living off crime, and varying degrees of organization may be employed to further their aims. Robbers and burglars, safe-breakers, and men who know how to dispose of stolen cars or valuable jewellery or antiques may have no means of support other than that which flows from crime. They too must properly be regarded as professional criminals, continuously employed in illegal activities when not serving prison sentences. Some have long records of previous convictions; some are more successful in avoiding detection; while others may have periods when they settle down, perhaps having married or co-habiting in a domestic environment which provides alternative satisfactions for a period of time.[18] Their motivation is part necessity (when in need of money), part opportunist (when the chance presents itself), and part habit (living up to a self-image).

Crimes of this sort produce sufficient returns for many criminals to live on fairly comfortably without resort to sophisti-cated laundering operations. An intermediary, in the shape of a fence, is the normal channel for converting the non-monetary proceeds of crime into disposable cash. 'Craft crime' refers, therefore, to those incidents where relatively small amounts of property or cash are stolen from a large number of separate victims in a routine fashion by essentially similar criminal groups. Offending of this sort may have the character of a

---

[17] *76th Report of the Commissioners of HM Customs and Excise* (Cmnd. 9655), HMSO, London, 1986, p. 27.

[18] See M. Maguire and T. Bennett, *Burglary in a Dwelling: The Offence, the Offender, the Victim* (Cambridge Studies in Criminology, Vol. XLIX), Heinemann, London, 1982.

collective enterprise, planned in advance although often poorly organized; and it is susceptible to social, community, and police remedies in a way that organized project crime is not.[19]

The containment of organized crime, whether falling into the project or the craft category as outlined above, is of fundamental importance to a healthy society. Although it is easy to fictionalize and glamorize such criminal activities, the reality is that they depend on cupidity, brutality, dishonesty, and immorality. Individual victims are harmed—physically if they stand in the way of the attainment of an objective (like the driver in the Great Train Robbery); financially if they are robbed or defrauded of their money or property. The victims apart, society as a whole is weakened and coarsened if organized crime is tolerated. Protection of the poor by the self-appointed, as in American urban folklore, can and does lead to extortion, exploitation, and graft. Corruption, actual or attempted, inevitably follows close on the heels of any acceptance of illegal activities, however tacit the acceptance, however discreet the illegality.

What should be the national response to the menace of organized crime which is so insidious and pervasive, which has such far-reaching economic impact, and which causes so much damage to the individual citizen? There are five stages, sequential but interrelated: prevention, detection, prosecution, conviction, and punishment. A discussion of prevention, a large and increasingly topical subject, is deferred until a later chapter. Of the other responses, after the event, detection, the result of successful policing, is the first line of defence against non-individual crime. Unlike domestic crimes, or incidents of affray such as the all-night drinking party mentioned at the start of this chapter, where the police may know nothing of a spontaneous incident until after it has occurred, organized crime is more of an evenly matched battle of wits. The police need not just material resources but also the intelligence, the imagination, and the capacity to combat adversaries who possess similar qualities but are not hampered by legal or political constraints. Outside the Metropolitan Police area, a high degree of specialist police manpower is concentrated in regional crime squads, linked to a

[19] See McIntosh, *The Organization of Crime*, p. 28; also J. A. Mack in collaboration with Hans-Jurgen Kerner, *The Crime Industry*, Saxon House/ Lexington Books, Farnborough, Hants, 1975.

national computer network, in each of the police regions of England and Wales. Criminal intelligence—the starting point for investigation, detection, and arrest—is gathered and collated on a regional basis; operations are carried out across the territorial boundaries of forty-three separate police forces; and inter-regional co-operation is developing as more and more protracted investigations are found to have national, or even international, connotations. The professional or 'target' criminals investigated by regional crime squads include armed robbers, major drug traffickers and operators of illicit drug factories, skilled forgers of banknotes and securities, and those organizing and perpet-rating large-scale thefts and burglaries.[20]

In the same way as a military campaign, police operations against organized crime cannot be conducted successfully without intelligence about the identity and intentions of the enemy. Writing about the history of warfare, Montgomery asserted that 'A general must understand the mind of his opponent, or at least try to do so.'[21] He added that for this reason he had always kept in his wartime caravan a picture or photograph of his principal opponent; thus, the face of Rommel accompanied him from the Western Desert to Normandy. Photo-graphs of police 'targets' are sometimes similarly displayed and studied. More practical information is hard to come by. Criminals and police officers inhabit different worlds, and whenever in the past attempts have been made by police to get closer to their adversaries, for example by socializing with the criminal fraternity in clubs and meeting places, the consequences in terms of contagion and corruption have outweighed any benefits in obtaining information leading to arrests. Moreover, today's first-division criminals keep a low profile, inhabiting a largely anonymous and mobile environment. Unlike the gangland bosses of twenty-five years ago, they do not need to strut about flaunting their wealth and power. Most take care to avoid the centre of the stage and emerge into the light of publicity only when something goes wrong.

[20] *Report of HM Chief Inspector of Constabulary, 1983* (HC 528), HMSO, London, 1984, p. 30.
[21] Field Marshal Viscount Montgomery of Alamein, *A History of Warfare*, Collins, London, 1968, p. 17.

Occasionally the police may catch a professional criminal red-handed, but more frequently an arrest will follow on information received. The sources of information are manifold, and sometimes murky. Observation and surveillance, intuition, and the interpretation of known facts are all valuable sources. Investigating officers also need to keep their ears close to the ground. Contrary to public belief, there is little honour among thieves: self-interest will normally prevail. Co-accused defendants and former associates may talk; others already convicted and serving prison sentences may have scores to settle or a wish to retire. Hence the role of the informer or 'grass', and his elevation in recent years to the status of 'supergrass'.

IV

Whereas police informers in the past have usually been anonymous, being paid or otherwise rewarded, the resident informer or supergrass is not so favourably placed. He is typically a convicted and sentenced offender, who, having been publicly tried in his own name for crimes he has committed, is prepared to give evidence, again in his own name, against his former associates. An effectual supergrass needs a particular aptitude for the accurate recall of details of past offences, as well as a willingness to appear in court to give evidence for the prosecution. In the main, although not exclusively, such informants have been found among the ranks of professional criminals specializing in organized robberies, a type of offence that has greatly expanded over the last three decades, with banks, post offices, and building society branch offices as the primary targets.

Places where currency has been stored in large quantities have attracted law-breakers ever since money became the medium for cash transactions. The first long phase of attacks on banks by professional robbers lasted from the nineteenth century until the early 1950s. The normal approach of the criminal was to enter the premises at night and to concentrate on areas of storage. The craft skill required was that of the safebreaker who, often when frustrated by improved locking systems, resorted to dynamite or other explosives to blow open safe doors. Greater ingenuity and higher technical standards of security provided more protection, but the advances in technology were not all on one side: safe-blowers turned to new cutting devices such as thermic lances. However, as physical security systems became increasingly effective in preventing criminals from reaching those areas of the bank where the cash was stored, a shift in approach became apparent.

In May 1952, an organized robbery which took place in the early hours of one morning ushered in a new era. The location was the West End of London, where seven men disguised with scarves and armed with coshes attacked a Post Office mail van and seized a total of £240,000 in used bank notes. The siren alarm on the van failed to operate: the Postmaster-General later had to admit to Parliament that it had not been tested as there was no way of turning it off.[22] From that time on, there was a clear trend away from trying to break into the near-impregnable strong rooms of banks and towards attacks by organized groups on the vehicles used to transport money from branch to branch, or between banks and business premises. Attempts to steal large amounts of cash in transit were supplemented by robberies, in which real or replica firearms were carried, and sometimes used, across the counters of banks, post offices, and building society branches during normal working hours. Members of the public might be present; if so, together with the staff, they would be threatened by the robbers. A very substantial wages robbery at London Airport on 27 November 1962 and the Great Train Robbery on 8 August 1963 were the most notorious examples of the scale of organized crime and the methods deployed. By the summer of 1972, armed robberies in the Metropolitan Police district were running at the rate of one in every five days, with the amount stolen in Greater London alone totalling some £3 million in the three years 1969–72. In the latter year—in August 1972—the speed and precision attained by organized bank robbers was displayed in an incident at a bank branch in Wembley, when £138,111 in bank notes was stolen in less than one minute.

Just before Christmas of that year, on 23 December 1972 at 5.30 pm, acting on information, Flying Squad officers travelled to Rushden, Northampton, and arrested Derek Creighton Smalls,

---

[22] *Parl. Debates*, HL, 176 (5th ser.), cols. 1287–9, 22 May 1952.

known to his friends as Bertie. He gave them no trouble, being drunk at the time, and offered no resistance. The following month, while in custody, Smalls indicated that he was ready to make a deal. In return for giving information about his erstwhile criminal associates, he asked for immunity from prosecution in a written guarantee—not just for himself, but for 'every robber in London'. Initially the police refused to listen; Smalls was by no means the first inveterate offender to offer to grass on his former colleagues. It was only during committal proceedings, after Smalls had been served with papers containing evidence of the charges against him, that he renewed his request, this time through a solicitor. A meeting followed between the solicitor and representatives of the police and Director of Public Prosecutions (DPP).[23] By now, the scale of what was on offer became apparent, and a judgment had to be made as to the morality and expediency of the response. The result was a document effectively granting Smalls immunity from prosecution. His expansive gesture towards every criminal in London was not pursued. In return, Smalls confessed to fifteen robberies and gave details of seven others. He named thirty-two criminal associates, with abundant supporting details. A majority of those arrested and charged were later convicted. No criminal had ever informed on such a scale before: the age of the supergrass had begun.

The DPP's decision not to proceed with prosecution in exchange for information of undoubted value to the police was strongly criticized. Defence counsel for Smalls' associates (understandably, in the circumstances of his clients) called it 'an unholy bargain'. The Court of Appeal was critical of the indemnity undertaking, although rejecting the appeals against conviction by sixteen of Smalls' accomplices. In giving judgment, Lord Justice Lawton was emphatic:

The spectacle of the Director recording in writing, at the behest of a criminal like Smalls, his undertaking to give immunity from further prosecutions, is one we find distasteful. Nothing of a similar kind must ever happen again. Undertakings of immunity from prosecution may have to be given in the public interest. They should never be given by the

[23] The Director of Public Prosecutions is a public official, answerable to the Attorney General, who is required by law to institute, undertake, or carry on criminal proceedings in any case which appears to him to be of importance or difficulty, or which for any other reason requires his intervention.

police. The Director should give them most sparingly; and in cases involving grave crimes it would be prudent of him to consult the Law Officers before making any promises.[24]

The House of Lords, in refusing leave to appeal on the point of law that an accomplice who has been promised immunity by the Crown should be prohibited from acting as a prosecution witness, was less exercised. Ironically, it was Lord Justice Lawton who came in for some unexpected criticism for being too outspoken in saying how the DPP, an official responsible to the Attorney General rather than to the judiciary, should go about his business.[25] In both courts the rule of law was reiterated that, while there was no bar to an accomplice giving evidence against his former associates, the judge must warn the jury as to the dangers of convicting without corroboration. There is no record of the effect of these proceedings on subsequent prosecution practice, although the Assistant Commissioner (Crime) at New Scotland Yard was later to record in his memoirs that it was agreed 'complete immunity would never be given again'.[26] From that time onwards, however, all the indications are that the DPP used his power to grant immunity as sparingly as the Court of Appeal had suggested that he should. No hard and fast rules were laid down, and little was said in public beyond a cautious affirmation that every case should be considered on its individual merits.

There are deep issues of public policy involved in any decision not to prosecute a professional criminal; and, although it has never been a principle of English law (as it is in some other countries) that, where evidence discloses the commission of an offence by an individual, then he *must* be prosecuted, nevertheless, the policy in practice has been that the more important the crime, the greater the public interest that proceedings should be instituted. Sometimes it will not be a straightforward decision. An individual may be informed that any statement he may make will not be used in evidence against him, but that no assurance can be given that he will not be prosecuted should other cogent evidence emerge at a later date. Nor would such an immunity

[24] *R.* v. *Turner and others* (1975) 61 Cr. App. R. 67 at p. 80.

[25] *DPP* v. *Brown and others.* Brown, like Turner, was one of the group of defendants convicted on Smalls' evidence. The case heard by the Appeal Committee of the House of Lords on 23 June 1975 was not reported, but a transcript exists of the proceedings.

[26] Kelland, *Crime in London*, p. 221.

extend to any perjury committed by him at a subsequent trial. There may also be a case for granting immunity where the evidence is insufficient to warrant the prosecution of an individual, but where there is reason to believe that he is in possession of information, possibly relating to organized crime or national security, which would be of value to the relevant public authorities. Certain informants may already be serving lengthy terms of imprisonment for other offences when they volunteer, for reasons of their own, to give evidence in court against former associates. If it seems unlikely that a consecutive sentence would be imposed, the DPP might decide, on the merits of the case, that the public interest would not be served by initiating a further prosecution.

Returning to the aftermath of the Smalls case, despite some suggestion that Smalls' life was in danger, it being stated in court that a very large sum of money had been offered for his 'execution',[27] informers continued to come forward. The information they were able to provide was clearly highly relevant to combating the epidemic of armed robberies. Sometimes the desire for vengeance played a part. In June 1974 Maurice O'Mahoney was arrested and charged with robbery, attempted robbery, and burglary. He had been arrested together with three other men, all of them older and more experienced in criminal offending than himself. While on remand, the three older men came to the conclusion (wrongly, as it transpired) that O'Mahoney had betrayed them. They attacked him, threatening to gouge out his eyes with a sharpened toothbrush. Their allegations riled O'Mahoney, and he resented their assault on him. In an application before the Magistrates' court, he told the court that he was guilty of the robbery of which he was accused and wished to inform the police about some other matters. His information resulted in numerous people being arrested and charged with serious offences, including conspiracy to murder and robbery. Property to the value of over half a million pounds was recovered. In this instance immunity from prosecution was neither sought nor granted, but it was not long before a further impetus was provided to encourage the supply of information from supergrasses, this time by the courts.

In 1977 a man named Charles Lowe, who had provided

[27] *The Times*, 7 Sept. 1973.

information to the police, was convicted and sentenced to eleven-and-a-half years' imprisonment. The Court of Appeal considered his appeal against sentence and reduced it by more than half. In substituting a five-year sentence, Lord Justice Roskill held that insufficient account had been taken of the enormous assistance rendered by Lowe to the police. In his judgment, he said:

It must be in the public interest that persons who have become involved in gang activities of this kind should be encouraged to give information to the police in order that others may be brought to justice and that, when such information is given and can be acted upon and, as here, has already been in part successfully acted upon, substantial credit should be given upon pleas of guilty especially in cases where there is no other evidence against the accused than the accused's own confession. Unless credit is given in such cases there is no encouragement for others to come forward and give information of invaluable assistance to society and the police which enables these criminals . . . to be brought to book.[28]

The defendant admitted to ninety-one offences (including fifteen robberies, some of them armed) and gave information which led to the arrest of thirty-seven other persons, with charges brought against twenty-four more, and the recovery of stolen property worth some £400,000. The police officer in charge of No. 5 Regional Crime Squad told the Court: 'Lowe's information is more important than any other information we have received from criminals of his standing ... As a result we are in the process of clearing up most of the serious gangs of criminals throughout the East End of London.'[29]

Lowe's case has subsequently been the basis of the advice given by solicitors to those of their clients who have been charged with the commission of serious offences and are prepared to plead guilty and assist the police. If, after informing and consulting with the DPP at an early stage, it is decided that their information is likely to be of value in investigating other crimes (particularly, although not exclusively, armed robberies), the police may apply to the court for such persons to be remanded with a condition of residence at a police station. In this way resident informants can be held in police custody during the sometimes lengthy period while information is being given and checked. But whether held by the police or on remand in

[28] (1978) 66 Cr. App. R 122 at p. 125.
[29] Ibid.

prison, an informant will face trial in the courts in the normal way and in his own name. Later it may be necessary for him to give evidence for the Crown in open court if convictions are to be obtained. All of this will inevitably put informants at risk of reprisals, and so, when released from whatever sentence they have served, they may be offered police protection and a new identity for their immediate family and themselves. But immunity from prosecution is seldom available.

V

Police practice has evolved over the last decade in parallel with policies on prosecutions by the Attorney General and the DPP, and sentencing in the courts. Operational changes in the Metropolitan Police have resulted in the Flying Squad being decentralized for working purposes with the task of investigating serious crimes. Outside London, the regional crime squads will liaise closely with Scotland Yard. The cult of personality has been strongly discouraged. The accent is now on a professional and impersonal team approach rather than the image of the ace detective or crime-buster so dear to the hearts of the crime reporters in the popular press. In the past, individual detectives, some of whom became household names and were very successful in bringing notorious criminals to trial, had made use of informers over the years in the pursuit of their investigations. But the dangers of unhealthy relationships developing, magnified by the corrupting effect of personal publicity, has led to a conscious change in policy.

Today's senior detectives may be less colourful, but they are more professional. They head teams of trained men who must operate within clearly defined police powers and are subject to constant public scrutiny. That is as it should be. Their task in obtaining information from a supergrass, if the informant is willing to supply it, is to test and check his recollection at every stage. Corroboration, in the form of independent evidence, is essential if convictions are to be obtained. The courts will rightly look with scepticism on the unsupported word of a convicted criminal giving evidence against his former associates. Judges are obliged to warn juries about the dangers of convicting on the evidence of co-conspirators, and the DPP will seldom feel justified in bringing charges on the basis of an informer's un-supported evidence. Charges may also be withdrawn if there is any evidence that in the course of their enquiries the police have offered an informer improper inducements. When in pre-trial custody, either at a police station (as a resident informant) or in prison, an informant must be treated in the same way as other prisoners awaiting trial, although he will need to be segregated (as are other inmates, such as child sex offenders) for his own protection. He must not be offered inducements in the form of more favourable treatment to persuade him to incriminate others. The same will apply if an informant has already been convicted and sentenced, but remains in police custody after conviction. In this event he must be subject to the same rules as would apply in a prison. Giving evidence for the Crown may be rewarded by a reduction in sentence, if the court so decides, but not by a more comfortable or privileged life while serving the sentence. Nor can early release on parole be promised by any police officer. To do so would not only be to offer a benefit that could not be honoured by the police, but would also amount to an inducement that could lead to charges being dropped against those incriminated by the supergrass.

Such men are often amoral individuals, with little regard for anyone except themselves. They can be cunning and devious, with personalities conditioned by a life of crime and (sometimes) punishment. Supergrasses are usually to be found among the ranks of the stronger personalities, and this needs to be matched in the qualities of the investigating officer. Both work side by side in building up the case against criminal associates, but it is of the utmost importance that officers do not become too closely involved emotionally with their charges. It is a relationship of captor and captive that says much about the contemporary world. When elated, an informant may consider himself part of an investigating team,[30] rejoicing in successes and downcast by setbacks. When depressed he may become moody, threatening, and uncooperative. Sometimes informants exhibit a genuine wish to put the past behind them and turn over a new leaf. They may have married, or may have young children growing up in the knowledge of their father's criminality. None of this can be put to

---

[30] Kelland, *Crime in London*, p. 222.

the test until the man is at liberty. As a further safeguard, the investigating officer will not have any part in the resettlement arrangements: these are the responsibility of another police department. At various stages in the enquiry, reference to more senior officers is mandatory.

The Metropolitan Police, with the regional crime squads, have been pioneers in resorting to resident informants in the campaign against organized crime. Taken together with other sources of criminal intelligence, informants probably constitute the most valuable single weapon at the disposal of the police. The results extend beyond the arrest and conviction of some of those responsible for the most serious and far-reaching crimes, leading to the disruption of organized groups, the recovery of weapons and the proceeds of crime, and the acquisition of further intelligence. Most important of all is the demoralization caused by a defector within a criminal organization. No one knows who can be trusted or who might go down the same path in the future. In his book *Crime in London*, Gilbert Kelland, who was Assistant Commissioner (Crime) at New Scotland Yard for seven years from 1977 to 1984, writes about the impact of resident informants on organized crime. Although no detailed statistics were kept, a count in 1982 showed that during a five-year period nearly fifty men had been used by the Metropolitan Police as resident informants, resulting in 451 people being charged, 262 being convicted, 85 being acquitted, 78 awaiting trial, and 32 cases not proceeded with.[31]

These are the reasons why, in considering how to respond to the methods of organized crime, the use of supergrasses cannot be ruled out. It may have an unsavoury taste, but the benefits to society are clear enough. Safeguards exist and need to be applied to regional crime squads and individual police forces outside the Metropolis. The 1984 Report of the Chief Inspector of Constabulary included a reference to the desirability of guidelines and raised the question of formulating a national policy towards the use of informants.[32] Such a development would not be breaking fresh ground. Two hundred years ago, Bentham eloquently pleaded for a more forthcoming attitude towards

[31] Ibid., p. 224.
[32] *Report of HM Chief Inspector of Constabulary, 1984* (HC, 469), HMSO, London, 1985, p. 31.

common informers. From Radzinowicz's *History of English Criminal Law*, we learn that Bentham sternly censured the attitude of the English law towards them. The law, he wrote, furnishes 'a curious but deplorable instance of human frailty. It employs them, oftentimes deceives them, and always holds them up to contempt.' He appealed to the public to change their attitude, and prophesied that such a change would bring innumerable advantages to society, and that the common informers themselves would respond by displaying 'high moral qualities'.[33]

Returning from this utilitarian vision to the present day, the last word on this prickly subject can be left with the Lord Chief Justice. In reducing the sentence on an informer on appeal from eight years to five, Lord Lane said:

It is not merely the value of this man as a witness which is important, although of course it is important; but possibly the true value of this sort of 'grassing', or informing on colleagues, is that it leads to other people doing the same thing ... One informer leads to another and, of course it is an estimable barrier of casting suspicion in the mind of one person in the criminal enterprise about the activities of his fellow men, an entirely desirable state of affairs. In the end it comes down to a question of public expedience. From the point of view of the public it simply is which is better: either to keep this man inside for 15 or 18 years, which is probably what his crime deserves, or to stimulate others to behave as this man did, in the hope thereby that less crime will be committed, and if it is committed more people will be caught and punished? In the view of this Court there is no doubt which of these two choices is better, the second one, the stimulation of the informer.[34]

In a subsequent case, in which the appellant had given information leading to the conviction of seventeen other offenders, the Lord Chief Justice laid down that the discount should be between one-half and two-thirds of the notionally appropriate sentence, depending on the value of the information given.[35]

---

[33] L. Radzinowicz, *A History of English Criminal Law and its Administration from 1750*, Vol. 3, Stevens, London, 1956, p. 436. Leon Radzinowicz was Wolfson Professor of Criminology at Cambridge, 1959–73, and Director of the Institute of Criminology, 1960–72. He was knighted in 1970.

[34] *Sinfield's* case, (1981) 3 Cr. App. R. (S) 258 at pp. 259–60.

[35] *R. v. King* [1985] Crim. L. R. 748 at p. 749.

VI

Since the prime object of organized crime is economic gain, it is to be expected that financial dealings will attract wrongdoers. Historically, as one of the authors in the Cambridge Studies in Criminology has pointed out, the control of financial and commercial fraud has been the Cinderella of the criminal justice system.[36] Compared with other crimes, fraud has until recently earned only limited police, prosecutorial, or media interest, despite the fact that the losses involved are far greater than those arising from theft, burglary, or robbery. In London alone, the cost of commercial fraud rose from £36 million in 1970 to £867 million in 1984. Fraud recorded by the Metropolitan and City of London Fraud Squad represented almost three times the total cost of all other property crimes in London, and was not far off the combined cost of reported thefts, burglaries, and robberies throughout England and Wales.[37] The climate has now changed, not just in Britain but in other advanced industrialized countries as well, partly as a consequence of pressures on governments to show that there is not one law for the rich and another for the poor,[38] and partly owing to a dawning realization of the sheer scale and ramifications of the largest frauds. The process has been accelerated by some well publicized financial scandals which have concentrated political and press interest on the City of London.

Unlike most other crimes, there is a problem of comprehension. Fraud is often elusive, hard to define, and hard to recognize. Most commercial transactions can be conducted honestly or dishonestly, and the dividing lines are not always clear. Financial frauds can be interspersed with legitimate trading activities or may be wholly fraudulent, such as VAT and revenue frauds. Banking frauds may involve the ingenious manipulation of accounts and duplicate accounts; marine frauds may lead to the same cargo being sold many times over by forged bills of lading and other documents. Phony companies may

[36] M. Levi, *The Phantom Capitalists* (Cambridge Studies in Criminology, Vol. XLIV), Heinemann, London, 1981.

[37] M. Levi, 'Some Criminal Justice Implications of a Survey of Fraud against Companies', Home Office Research and Planning Unit *Research Bulletin*, no. 21, 1986, p. 14.

[38] Levi, *Phantom Capitalists*, p. xiii.

obtain goods on credit from genuine suppliers, and phony suppliers may sell non-existent goods to genuine buyers. The difficulties of investigation are as great as bringing the cases before the courts and proving that criminal offences have been committed.

In England and Wales a milestone was reached in November 1983 when the Government announced an intention to establish a Fraud Trials Committee, under the chairmanship of a law lord, Lord Roskill,[39] to consider the ways in which the conduct of criminal proceedings arising from fraud could be improved, and to recommend changes in the existing law and procedure 'to secure the just, expeditious and economical disposal of such proceedings'.[40] The Committee was appointed in April 1984 and reported in January 1986, coincidentally with the consideration by the House of Commons of a Financial Services Bill. This legislation was aimed primarily at those areas of the securities and investment markets either where statutory regulation had not existed, or where it had failed to perform adequately in protecting investors. In the words of the chairman of the Stock Exchange, the Bill proposed 'the equivalent of store detectives on the inside of these industries with a state-appointed agency standing by to chastise them if they fail in their duties, or even to take over these duties if necessary'.[41] In introducing the Bill on Second Reading, the Secretary of State for Trade and Industry, Leon Brittan, said it did not extend to the Lloyd's insurance market (which is regulated by the Lloyd's Act 1982), but that he had decided to set up an 'independent, full and rigorous inquiry' into Lloyd's under the chairmanship of Sir Patrick Neill, a past chairman of the Council for the Securities Industry and the Vice-Chancellor of Oxford University.[42] Cinderella had arrived at the ball, and the spotlight was on her.

Much of the argument on the regulation of financial dealings has turned on whether direct statutory supervision and control is likely to provide for the better protection of investors than self-regulation by the institutions themselves. In fact, the two are

---

[39] Lord Justice Roskill, who had presided in the Court of Appeal in Lowe's case, became a Lord of Appeal in Ordinary in 1980. He retired in 1986.
[40] *Fraud Trials Committee Report*, p. 5.
[41] *The Times*, 16 Jan. 1986.
[42] *Parl. Debates*, HC, 89 (6th ser.), col. 946, 14 Jan. 1986.

closely related, and the framework now provided by the Financial Services Act 1986 combines aspects of both, requiring the authorization of a large number of investment businesses,[43] either directly via a designated agency (the Securities and Investments Board), exercising a transferred power from the Secretary of State, or indirectly via membership of one of a number of self-regulating organizations (SROs) or recognized professional bodies. The SROs are themselves subject to supervision by the Securities and Investments Board (SIB).

Twelve months after its appointment, the Neill Committee reported on the regulation of Lloyd's.[44] The committee resisted political pressure to bring Lloyd's within the ambit of the Financial Services Act, preferring a more subtle scheme of independent oversight tailored to the peculiar character of Lloyd's. Investor protection is not a concept that readily fits the insurance business as it is carried out at Lloyd's. Those who join underwriting syndicates, so becoming 'Names', are not investors in the generally understood sense: they are individual insurers with the legal status of sole traders whose liability is unlimited.[45] In joining Lloyd's, a Name puts his entire wealth at risk to provide insurance cover. In most respects it is the policy-holders insuring against risks at Lloyd's who are the investors, and not the Names who underwrite the risks. Policy-holders already have effective protection, and it was spectacular abuses by a few dishonest working members of Lloyd's, taking advantage of their Names, rather than any dissatisfaction expressed by policy-holders over the payment of claims, that led to the clamour for closer regulation and external supervision. The solution proposed by Sir Patrick Neill and his colleagues was to modify the balance on the Council (i.e. the governing body of Lloyd's) so that the whole-time working members ceased to have an overall majority. Besides the elected representatives of Names, it was envisaged that the number of nominated outside members, each requiring the prior approval of the governor of the Bank of England, should be doubled from four to eight. In order to institutionalize a link with the Securities and Investments Board,

[43] Numbering some 15,000 on the best available estimate in 1986.
[44] *Regulatory Arrangements at Lloyd's: Report of the Committee of Inquiry* (Cm. 59), HMSO, London, 1987.
[45] Ibid., pp. 5–6.

although remaining outside its sway, the chairman of the SIB for the time being (or his deputy) should be included among the nominated members of the Council. The Rules Committee should be chaired by a nominated member, and the nominated members should make their own report annually on the progress of regulation at Lloyd's.

None of the seventy recommendations called for legislation, and in a statement to the House of Commons on publication of the Report the Secretary of State for Trade and Industry said that the proposals were a good basis for the further development of measures to protect the interests of Lloyd's members, inferring that, provided agreement was reached for their early implementation to bring the standard of protection of members' interests up to that provided by the Financial Services Act, the recommendations were likely to be acceptable to the Government.[46] Although those who place their faith in the efficacy of statutory controls, sometimes without any understanding of how they would work in practice, may have been disappointed, the proposals offer a workable system of self-regulation, externally buttressed. As the Committee of Inquiry pointed out, additional experienced outsiders brought on to the Council as nominated members will perform internally a role not markedly different from that which members of any external supervisory authority could be expected to undertake if given the task of overseeing the effectiveness of regulation at Lloyd's.[47] The nominated members are also likely to be far better informed than any external body could hope to be about the byzantine working practices without their intervention being resented and circumvented.

The debate on statutory versus voluntary regulation in the financial sector has been a heated one, embracing ideological attitudes as well as empirical considerations. The advocates of self-regulation argue that practitioner-based controls have the advantage of drawing on first-hand knowledge of where and how corners can be cut. At the same time, it is now a statutory requirement for any authorized person to maintain a proper separation between the use of his own and his firm's money, and

[46] *Parl. Debates,* HC, 108 (6th ser.), col. 1035, 22 Jan. 1987.
[47] *Regulatory Arrangements at Lloyd's: Report of the Committee of Inquiry* (Cm. 59), pp. 82–3.

the use of money belonging to his clients which is held on trust. Although the City of London and other financial institutions do not share the same history and traditions as the older professions, there seems no good reason why they should not aspire to similar standards. For many years erring doctors, lawyers, and accountants have been subject to disciplinary proceedings by their respective professional bodies if they have fallen short of the conduct expected of them. The reputation of their professions has been regarded as demanding nothing less. Public confidence has been seen to be at stake. Sanctions, including expulsion or suspension, can be harsh, in the most extreme cases depriving members of their professional livelihood. Appeal procedures exist, and if there is evidence that a criminal offence has been committed, the individual may also be subject to prosecution in the courts.

Certain financial institutions, notably the Stock Exchange, have already moved some way in this direction. In January 1986 the Stock Exchange was reported to have 4,500 members grouped in just over two hundred firms. More than two hundred regulatory staff were employed, including lawyers, accountants, and former police officers. Twenty-five of these had the status of inspectors, with powers to scrutinize in detail all of a firm's records and to investigate any suspicion of irregularity or malpractice. Between 1978 and 1985 twelve members of the Stock Exchange were expelled, with twenty-three more suspended for lesser disciplinary offences, and a further twenty-three censured publicly. In a majority of cases the malpractices had been uncovered during visits by the Stock Exchange's own inspectors. The total annual cost of the surveillance division was running at half a million pounds in 1985.[48] 'Self-regulation', declared the then Secretary of State pithily in opening the Second Reading debate on the Financial Services Bill in January 1986, 'does not mean relying on people's consciences; it means judgement by one's peers.'[49]

The new measures to give better protection to investors or users of financial services, supplementing practitioners' standards of probity, are backed by criminal sanctions. In future, nobody can carry on an investment business unless he is authorized to do

[48] *The Times*, 16 Jan. 1986.
[49] *Parl. Debates*, HC, 89 (6th ser.), col. 943, 14 Jan. 1986.

so. The penalty for contravention is up to two years' imprisonment, or a fine, or both. Increased protection is called for if the consumer (possibly investing for the first time and having been attracted by paid publicity or salesmanship) is not to suffer as a result of the greatly increased competitiveness in the City of London and other markets. More effective machinery is also needed to detect and deter deliberate and organized fraud. In commenting on the enforcement of the law, the Fraud Trials Committee did not mince its words:

The public no longer believes that the legal system in England and Wales is capable of bringing the perpetrators of serious frauds expeditiously and effectively to book. The overwhelming weight of the evidence laid before us suggests that the public is right. In relation to such crimes, and to the skilful and determined criminals who commit them, the present legal system is archaic, cumbersome and unreliable. At every stage, during investigation, preparation, committal, pre-trial review and trial, the present arrangements offer an open invitation to blatant delay and abuse. While petty frauds, clumsily committed, are likely to be detected and punished, it is all too likely that the largest and most cleverly executed crimes escape unpunished.[50]

The Committee advanced a large number of recommendations, 112 in all, and some of them radical, for changes not only in the substantive law but in the procedures to be followed between the time that a fraud first becomes known to the authorities and the moment the verdict is given in court. 'Some of our proposals may shock traditionalists', the Committee observed, adding: 'the same was probably true of the proposal to abolish the medieval practice of trial by combat.'[51] With this warning, the Report went on to make its single most controversial proposal: to abolish the right to a jury trial in the most complex fraud cases. Despite its shortcomings, the Committee concluded that trial by jury was an acceptable procedure in the generality of cases and should continue. That said, a majority of seven members to one went on to recommend a different type of tribunal, referred to as a Fraud Trials Tribunal, for the most complex cases falling within certain guidelines. Such a tribunal would consist either of a High Court or a circuit judge and two lay members selected from a panel of persons with experience of

[50] *Fraud Trials Committee Report*, p. 179.
[51] Ibid., p. 3.

business dealings and the capacity to comprehend the kind of issues that are likely to arise.

The majority on the Committee preferred this solution to resorting to special juries made up of people of above-average education, training, and experience. In rejecting that alternative, it was pointed out that, far from being an innovation, the requirement of special juries for fraud cases would put the clock back. Special juries had been an established feature of legal machinery in one form or another from the fourteenth century until their demise in the mid-twentieth century. The qualifications for a special juror were more exacting than those of a common juror: he had to be either an esquire or a person of higher rank, a banker or a merchant, or the occupier of a house with a rateable value higher than the level set for service as a common juror.[52] This was not the language of postwar egalitarianism, and special juries were abolished in 1949, except for certain commercial cases which could be tried by a special jury drawn from the City of London. In the event, the City of London Special Jury sat only three times after 1949 and was finally abolished in 1971. Quite apart from these historical considerations, the Roskill Committee ruled out a return to trial by special jury on the grounds that they did not believe that special jurors would have the degree of knowledge or expertise that would be required in order properly to grasp the points of concern in a complex case.[53] The Committee also made detailed proposals aimed at achieving greater investigative efficiency and closer co-operation between various agencies investigating fraud. The report envisaged the establishment of a Fraud Commission overseeing progress in fraud cases and the appointment of case controllers to be responsible for the control of individual cases of serious fraud from the time of discovery to the verdict in court.

One of the members of the Committee, Lord Benson, later gave a vivid description of the character of a complex fraud trial by drawing on his practical experience as a partner in one of the largest firms of chartered accountants:[54]

It would comprise over 100 witnesses, many of them of doubtful probity. There would be 5,000 exhibits chosen from many thousands more.

---

[52] Ibid., pp. 144–5.   [53] Ibid., p. 145.

[54] Lord Benson was a partner in Coopers and Lybrand (formerly Cooper Brothers and Co.), 1934–75, and was chairman of the Royal Commission on Legal Services, 1976–9.

There would be 50 or 60 banking accounts paid all over the world in different currencies. Many of the records could not be seen. They would be computer records and marks on discs. There would be a network of 50 or 60 companies spread round the world locking or interlocking in a way which it is almost impossible to discover. There would be six or seven accused with varying degrees of guilt or innocence. There would be six leading silks, each placing a different interpretation on the evidence of the oral witnesses and the documents which are put forward as exhibits. The trial would last from 30, 40 or 50 days—the longest we have had on record so far on the evidence is 137 days.[55]

In a statement to the House of Commons shortly after publication of the Report, the Home Secretary, Douglas Hurd, forecast legislation to implement many of the Roskill recommendations, saying that the Government was determined to bring about the changes in law, practice, and attitudes that were necessary to bring the perpetrators of serious fraud to book.[56] On the contentious issue of jury trial, he kept his powder dry, indicating that he would like to test opinion further before making up his mind.[57] Opposition spokesmen and government back-benchers combined to express doubts about the proposal. In the House of Lords the Lord Chancellor was more robust. Lord Hailsham said that he was not so completely convinced as other peers that the jury is an adequate safeguard in complicated cases for the acquittal of the innocent. It would be necessary 'to examine our whole list of sacred cows and consider which of them in fraud cases we are going to retain for all purposes'.[58]

VII

Sacred cow or pillar of the constitution? The question has appeared in papers set for law students for generations, and the

---

[55] *Parl. Debates*, HL, 471 (5th ser.), col. 40, 10 Feb. 1986.

[56] *Parl. Debates*, HC, 89 (6th ser.), col. 925, 14 Jan. 1986.

[57] A White Paper, titled *Criminal Justice: Plans for Legislation* (Cmnd. 9658), which followed in March 1986, accepted the Roskill Report as 'an excellent basis for legislation and administrative action' and summarized the arguments for and against the proposed Fraud Trials Tribunal. It did not, however, reach any conclusion, simply undertaking to consult urgently on the proposal (pp. 17–18).

[58] *Parl. Debates*, HL, 469 (5th ser.), col 986, 14 Jan. 1986.

wider debate has always evoked strong passions. The dissenting member of the Roskill Committee, Walter Merricks, remarked with some justification that if Britain had a written constitution one of its first articles would declare that, in criminal cases in which the accused was liable to be sentenced to a long prison sentence, he should have the right to trial by jury. It was a right that citizens had enjoyed for centuries, and previous attempts to remove or tamper with it had been robustly resisted. The Government should not be panicked into trying again because of hysteria over certain well publicized frauds.[59]

Coolly assessed, where does the balance of advantage lie? The agreed objective must be to counter the evil of widespread fraud in the most effective way. Unlike the use of supergrasses by the police, contained within the normal procedures of prosecution and trial, the Roskill proposal embodied a significant departure from principle. Yet the Committee's own report revealed that, in 1984, 86.8% of those brought before the Crown Court for offences of fraud were convicted,[60] a total that slightly exceeded the figure of about 85% of those convicted of all offences in the Crown Court in the same year. Nor, according to Merricks, was there evidence from the prosecuting authorities that charges were not laid through fear that juries would not be able to grasp all of the complexities.[61] Inadequate resources for investigation are more often to blame when frauds are not pursued. Moreover, how is complexity to be determined? Where would the dividing line be drawn? No doubt it is tedious and vexing for those with greater knowledge or faster minds, but it is one of the distinctive features of the jury system that the prosecution must simplify the facts, just as the judge must simplify the law, in a manner that can be understood by men and women of ordinary education. In this way an accused person can be sent to prison or otherwise sentenced only after a trial in which the central issues are comprehensible to all. If the law and the proceedings in court are too complicated for most citizens to understand, then the legitimacy of the courts will surely be at risk. In a felicitous phrase, Lord Devlin once described each jury as a little parliament, a place for discussion and debate between ordinary people

---

[59] *The Times*, 14 Jan. 1986.
[60] *Fraud Trials Committee Report*, Appendix K at p. 243.
[61] *The Times*, 14 Jan. 1986.

before a decision is reached. He went on: 'The jury sense is the Parliamentary sense. . . . trial by jury is more than an instrument of justice and more than one wheel of the constitution: it is the lamp that shows that freedom lives.'[62] Although my purpose in this chapter is to argue in favour of measures to combat organized crime, I am not persuaded by the case that has been made for abandoning the jury in this instance.

In the event, it may turn out that some of the less controversial proposals made by the Roskill Committee relating to evidence and criminal procedure may have as much or more impact on the conduct of fraud trials than any change in the constitution of the court. There is much to be said for revising rules of evidence to allow more documents to 'speak for themselves' without requiring them to be supported by witnesses. In its report, the Committee referred to the necessity for substantial revision in order to deal with the international criminal and the deliberate obstructionist.[63] Other practical steps advocated by Roskill and accepted by the Government included enabling the courts in England and Wales to request authorities overseas to take evidence on commission for use in this country, or for documents to be produced; and for a preparatory stage of the trial to be held at which points of law could be settled and the issues in dispute clarified before the jury is sworn in. Harnessing modern technology would enable witnesses as far afield as New York or Sydney to 'appear' via satellite video links on screen in the courtroom and to be examined and cross-examined by counsel without any loss of immediacy.[64] A clause in the Criminal Justice Bill 1986/7 proposes that witnesses outside the United Kingdom should be allowed to give evidence on a live video link, and also that children under the age of fourteen be permitted to give evidence by this means in respect of certain assault and sexual offences.

The Government is committed to strengthening the ways in which complex fraud cases are investigated, as well as prosecuted, and an overall co-ordinating body of some kind is essential if the present fragmentation is to be remedied. In its

[62] *Trial by Jury* (rev. edn) (Eighth series of Hamlyn Lectures), University Paperback, Methuen, London, 1966, p. 164.

[63] *Fraud Trials Committee Report*, p. 179.

[64] Ibid., p. 76.

first recommendation, the Roskill Committee proposed a single unified organization to be responsible for all the functions of detection, investigation, and prosecution of serious fraud cases.[65] After a study by the Chief Secretary to the Treasury and other interested Ministers, the Government came forward with the idea of a Serious Fraud Office, separate from the police but collaborating with them in the investigation and prosecution of the most serious and complex frauds. This new statutory body would have some, but not all, of the characteristics sought by Roskill. It is envisaged that the staff will have a range of relevant experience and qualifications, including those of lawyers, accountants, and people familiar with companies legislation and the commercial world. Some will be seconded from other parts of Whitehall, but the intention is to cast the net sufficiently wide to include people from the private sector as well. While the objective is to combine the necessary powers and expertise to enable major fraud cases to be brought to trial as rapidly and effectively as possible, this will happen only if the deeply engrained institutional reluctance on the part of the existing departments and enforcement agencies can be overcome to permit them to pull together. The ability to move quickly is crucial if the public is not to become disillusioned. Going over the same ground twice, as can happen when police enquiries follow a Department of Trade and Industry inspectors' report, is time-consuming and can cause delays in bringing cases to trial where there is evidence to justify a prosecution. A separate Roskill proposal for an independent monitoring body, described as a Fraud Commission, to observe the progress of fraud cases through the courts, advising on the time taken and the causes of delay, was not pursued on grounds that the benefits would be unlikely to outweigh the costs.

On the most controversial matter of all, having taken the temperature inside and outside Parliament, the Home Secretary decided against curtailing the right of trial by jury in complex fraud cases. When published in November 1986, the Criminal Justice Bill omitted the proposal on which a majority of the Fraud Trials Committee had set such store. But it is likely that the argument will continue, since many people remain convinced

[65] Ibid., p. 179.

that special tribunals are needed to unravel and resolve the most complicated financial frauds. When the House of Lords debated the report of the Fraud Trials Committee in February 1986, Lord Roskill's argument in favour of a non-jury tribunal was supported by two present and two past Lords of Appeal in Ordinary.[66] They included Lord Denning, who pronounced: 'I am afraid we have to leave trial by jury and we have to replace it by a modern instrument.'[67] The issue is by no means yet dead. In his Second Reading speech in the House of Commons, the Home Secretary said that he did not rule out the possibility of returning some time to the idea of a special tribunal, but that he would not be pursuing it in the Criminal Justice Bill.[68]

In parallel with these procedural responses are measures to deprive offenders of the profits of their crimes. Ever since 1978, when the Crown Court at Bristol ordered the forfeiture of substantial assets in the hands of a criminal conspiracy which had profited hugely from the manufacture and sale of the hallucinogenic drug known as LSD (an order that was later overturned in the House of Lords on appeal), there has been mounting interest in ways of depriving convicted offenders of their ill-gotten gains. In the Operation Julie drugs case, the prosecution was able to trace some £750,000 of profits from trafficking in LSD in the hands of those convicted. The assets included cash, cars, deposits of money and securities held in bank accounts in Switzerland and France, paintings, and electrical equipment.[69] While sympathetic to the aims of the court, the House of Lords decided that orders of forfeiture could apply only to tangible objects directly related to the illegal activity, such as the drugs involved, the apparatus for making them, the vehicles used for transporting them, or cash ready to be (or having just been) handed over in payment for them. The House also held that an English court had no jurisdiction to make orders for the transfer of property abroad.[70]

---

[66] Lords Griffiths, Templeman, Wilberforce, and Denning. *Parl. Debates*, HL, 471 (5th ser.), cols. 7–73, 10 Feb. 1986.

[67] Ibid., col. 55.

[68] *Parl. Debates*, HC, 106 (6th ser.), col. 466, 27 Nov. 1986.

[69] *Profits of Crime and their Recovery: Report of a Committee chaired by Sir Derek Hodgson* (Cambridge Studies in Criminology, Vol. LII), Heinemann, London, 1984, p. 3.

[70] *R. v. Cuthbertson and others* [1981], AC 470.

This judgment led to the setting up of a committee under the auspices of the Howard League for Penal Reform, with two Queen's Bench Division judges as well as a chartered accountant and a solicitor, included in its membership. The scope of the committee's enquiries extended to compensation and restitution of property for the victims of crime, and this aspect was referred to in the previous chapter. In its report, published in 1984, the committee recommended that the Crown Court should be empowered to order the confiscation of the proceeds of an offence of which the victim had been convicted or had asked to be taken into consideration. There should be a prescribed minimum, but no maximum limit. Normally the burden of proving the amount of gross receipts would be on the Crown, but where a defendant was convicted of the wholesale supply of class A or B dangerous drugs with a street value of £100,000 or more, he would have to prove that any assets acquired after the date of his first offence were legitimately obtained.[71] The committee also confirmed that the use made of the criminal bankruptcy powers given to the courts in the Criminal Justice Act 1972 had been largely ineffective.

These recommendations were generally well received, being welcomed in Parliament, and new powers to confiscate the proceeds of trafficking in controlled drugs were incorporated in the Drug Trafficking Offences Act 1986.[72] This statute confers powers, and obligations, on the Crown Court to confiscate the proceeds even after they have been converted from cash into some other type of asset; it also restrains a defendant from disposing of property before the end of his trial. In certain circumstances property transferred to other people is also liable to confiscation. Once convicted, the court will assume that the whole of an offender's property, together with any assets that may have passed through his hands in the previous five years, represents the proceeds of trafficking in drugs. There are no lower limits as recommended in the Hodgson report. The onus of proof is on the convicted person to show which if any of his

---

[71] *Profits of Crime and their Recovery,* p. 151–2.

[72] Drug trafficking is not in itself an offence known to law. It is a description of certain illegal activities including the production, transport, storing, or supply of controlled drugs; their improper import or export or the fraudulent evasion of Customs regulations; and the new offence (introduced in section 24 of the 1986 Act) of assisting another person to retain the benefit of drug trafficking.

assets were legitimately acquired. This rebuttable assumption represents a reversal of the normal onus of proof, although coming into play only after conviction. On the substantive drug trafficking charges, it will still be up to the prosecution to prove the guilt of the accused person beyond reasonable doubt in the normal way. Thus, it is a defendant's property rather than his liberty that is at stake. The restriction to Crown Court cases should exclude most of those addicts who supply drugs to their friends and others on a small scale to finance their own addiction, since a majority of such cases are dealt with in the Magistrates' court.

Confiscation represented a radical response to the special problems raised by drug trafficking. In the following session of Parliament, the Criminal Justice Bill proposed the extension to other profitable crimes of a power enabling the courts to order confiscation of the proceeds of offences from which at least £10,000 had been made.[73] Confiscation will relate only to the proceeds of offences charged or taken into consideration for sentencing purposes. Unlike the Drug Trafficking Offences Act, there will be no assumption that all of the assets at the offender's disposal were the proceeds of crime. A further limitation is that the court will have discretion over whether or not to make an order and for how much, rather than being required to order the confiscation of the full amount of the proceeds. Once a confiscation order has been made, however, any of the defendant's available assets can be realized to satisfy the amount imposed, and if he refuses to give up known assets he can be sentenced to prison in default, in addition to the term imposed as a penalty for his offences.

It can be argued that confiscation is an additional penalty over and above those authorized by Parliament. Libertarians, on the right as well as the left, have contended that an exceptional power introduced to deal with an exceptional threat posed by drug trafficking is being unwarrantably and hastily extended to a wide range of other offences. Few voices are more eloquent than

[73] The power to confiscate the proceeds of an offence will generally be limited to the Crown Court, although the Criminal Justice Bill proposes that Magistrates' courts should exercise the power in cases of offences relating to sex establishments and infringements of the Video Recordings Act 1984 or the Cinemas Act 1985. Other offences may be added by statutory order.

Enoch Powell, who opposed the principle of confiscation from the start, but let what he described as a 'revolutionary innovation' go by on the reasoning that drug trafficking was so dangerous, so abhorrent, and such a threat to society that it justified an unprecedented response. But now, at the first opportunity, he claimed, the precedent was being repeated.[74] There is force in this argument, and yet the need for a further sanction that may be expected to bite on those who profit from large-scale organized crime is evident. Criminal bankruptcy was tried before, but proved ineffective. Moreover, if deterrence bears only on a calculation of the costs and benefits of deliberately planned crimes, as some criminologists believe, confiscation could be a potent factor in adjusting the balance to the advantage of society as a whole. In terms of social justice, too, there are questions to be asked and answered about the propriety of leaving with a convicted offender the proceeds of crimes that belong not to him, but to others. Is it right that, having served his sentence, an offender should live comfortably on the fruits of his illegal and anti-social conduct? With the safeguards contained in the Bill, and any others that Parliament may choose to add during its legislative passage—for example, on restoration or compensation in the event of wrongful or erroneous confiscation—the power of confiscation represents a strengthening of what should be (although is not always) a process of criminal justice that is sensitive to changing situations.

VIII

Obscure, complex, and elusive as it may be, the flourishing condition of organized crime is an affront to any civilized society. It is not a new manifestation; England has had a criminal underworld, crudely organized, for as long as it has had any formal process of criminal justice. But the scale and sophistication have greatly increased, and all the indications point towards continuing growth rather than decline. The outward forms have changed with the disappearance from the scene of the well publicized London gangland bosses of the 1960s. Today the magnitude of organized crime only occasionally emerges into the

[74] *Parl. Debates,* HC, 106 (6th ser.), cols. 488–9, 27 Nov. 1986.

light when a case comes before the courts, although the principal organizers frequently avoid prosecution. It is not often that there is sufficient evidence to bring charges against them, even if the identity of the ringleaders has been established in the course of an investigation.

Organized crime cannot be shrugged off as something that is inevitable in the modern world—regrettable, of course, and heartily to be condemned, but too costly and difficult to get to grips with. Such toleration is more than just taking the easy way out, abrogating responsibility: it is to allow the gradual spread of creeping corruption and the undermining of standards of public morality. The many examples to be found in the United States are a warning. What is imperative is to mount a determined and co-ordinated national campaign, mobilizing all of the relevant agencies—not only the police, but the Inland Revenue, Customs and Excise, the Department of Trade and Industry, and other public bodies, supported by the private sector especially in instances of financial fraud. The police need not feel that their investigating role and expertise will be compromised, but rather that, with their help, investigations will be sharpened and made more effective.

There are already moves in this direction, and they need to be intensified and maintained. Publicity is a spur. Investigative reporting in the press and broadcast media is seldom welcome to those whose activities are being probed. The revelations that result are frequently lacking in accuracy in every detail, but they can sound the alert when something is wrong. Without publicity, the uninformed public cannot be expected to be aware of the nature or extent of organized crime, or of what can be done to assist in its prevention and control. The provision of information by private individuals is one practical response. At a deeper level, a more widespread and deliberate adherence to standards of probity in public and commercial life can make the water colder for the wrongdoer. Anyone who shies away from this admonition as too high-minded and unworldly need look no further than the report of Lord Salmon's Royal Commission on Standards of Conduct in Public Life, which followed in the wake of the Poulson affair in the early 1970s:

Neither ... rules nor codes of conduct can make people honest or be any substitute for an individual's sense of right and wrong. In the last resort an individual can be guided only by his common sense and his moral sensitivity; the test is whether he judges that his behaviour could withstand public scrutiny.[75]

The resort to supergrasses, the proposals to vary the normal rules of evidence and procedure in complex fraud trials, and legislation to deprive certain convicted offenders of their assets and property are all exceptional measures. They represent a response, an expedient response, to the threats posed to the community as a whole by organized crime, particularly robbery, drugs trafficking, and fraud. None of them lies easy on the conscience, and it is the responsibility of Parliament and the courts to assess the implications for civil liberties. The question for the public is whether such changes in the law and its enforcement outweigh the consequent diminution in the liberties of the individual citizen. In so far as they redress an imbalance, with the probability that more of those engaged in organized crime will be detected and prosecuted, and that more of those prosecuted will be convicted, and that more of those convicted will be discouraged by the sentence received and orders made by the court from returning to their former ways on release, then, with the one reservation relating to trial by jury, I conclude, however reluctantly, that measures of this sort are justified. To exclude expediency from criminal justice altogether, as purists might wish, would be to make it less responsive to the influences and practicalities that surge around it, shaping it and, ultimately, upholding it.

[75] *Report of the Royal Commission on Standards of Conduct in Private Life* Cmnd. 6524), HMSO, London, 1976, p. 64.

4

Mentally Disordered Offenders: Remedial or Penal Responses?

I

The professional criminal making a living out of deliberately organized crime and the mentally disordered offender stand at opposite ends of the spectrum of crime. In spite of the fact that the term 'mental disorder' has enjoyed statutory definition since 1959, the description 'mentally disordered offender' remains obstinately imprecise. According to Section 1 of the Mental Health Act 1983 (which consolidated the earlier legislation), mental disorder means mental illness, arrested or incomplete development of mind, psychopathic disorder, and any other disorder or disability of mind.[1] In non-specialist language, this falls well short of a specific definition including a variety of conditions, ranging from mentally retarded or inadequate people who find it hard to cope with the demands of everyday life, through various forms of persistent disorder or disability, to those suffering from acute mental illness (sometimes of a temporary nature) or impairment associated with abnormally aggressive or seriously irresponsible conduct. Far more offenders are inadequate or disordered than are mentally ill or severely impaired, but it does not follow that the degree of mental disturbance is reflected in the gravity of the offending or the response of the penal system. Very serious crimes of violence up to and including wounding and homicide can be committed by normally harmless, and sometimes pathetic, people who suddenly lose their self-control.

Mental disorder and abnormality of mind are very nearly, but not quite, the same thing. Diagnostically, either term will often suffice to describe an offender's mental state. But to the Royal

[1] Mental disorder was defined in these terms in Section 4(1) of the Mental Health Act 1959.

Commission on the Law Relating to Mental Illness and Mental Deficiency in the 1950s, 'abnormality' implied a permanent condition, whereas 'disorders', which were susceptible to treatment, might be temporary.[2] The Committee on Mentally Abnormal Offenders under the chairmanship of Lord Butler, which followed in 1975, disowned its title by avoiding the term 'mental abnormality' throughout its report.[3] The Committee cited people who could be said to be mentally abnormal in the sense of diverging from the statistical norm of mental functioning, although not necessarily being disordered. The examples they gave were religious fanatics, occasional drunks, and people with unusual mental capacities such as photographic memories. Yet abnormality of mind also has a place on the statute book, being the basis of the all-important defence of diminished responsibility in homicide cases.[4]

In the commentary that follows, it will soon be apparent that the lawyers' pursuit of exactness in meaning is not always compatible with medical phraseology and the classification for treatment purposes of mental patients, some of whom defy agreed diagnosis. The absence of common ground is the cause of much misunderstanding and mutual mistrust. The risk is that, as a result of professional disagreements, offenders may end up in the wrong type of institution. An example of lack of precision can be found in the first section of all of the Mental Health Act 1983. After describing the scope of the Act and saying what is meant by the term 'mental disorder', Section 1(3) states that no person should be regarded as suffering from a mental disorder for the purpose of the Act by reason only of promiscuity or other immoral conduct, sexual deviancy (this has been interpreted to include homosexuality), or dependence on alcohol or drugs. Ambiguity arises from the fact that, only too evidently, the prolonged abuse of drugs and alcohol can contribute to mental disorders. These exclusions do, however, as is the intention, provide a useful safeguard against locking away in hospital those unfortunate but not criminal people whose alcoholism, drug

---

[2] *Royal Commission on the Law Relating to Mental Illness and Mental Deficiency* (Cmnd. 169), HMSO, London, 1957, p. 24.

[3] *Report of the Committee on Mentally Abnormal Offenders* (Cmnd. 6244), HMSO, London, 1975, p. 4.

[4] Homicide Act 1957, S. 2(1).

dependence, or sexual misbehaviour has become an embarrass-
ment or a nuisance. But the borderline is a fine one. It is for this
reason that a discussion on drugs-related offending, which might
otherwise form part of this chapter, is postponed until later.

While the legislation on mental health applies to mentally
disordered persons generally, and sets out the conditions that
must be fulfilled to allow compulsory detention in hospital for
assessment or treatment, there are special provisions affecting
those who have been charged with a criminal offence. If found
guilty, they will then be dealt with either under the mental health
powers of the state (by compulsory reference to a hospital for
treatment),[5] or by recourse to penal powers (resulting in a
custodial or non-custodial sentence). These are the only two
grounds for depriving the citizen of his liberty, either for a fixed
period of time or indefinitely. A separate power of temporary
detention allows the police to remove apparently disordered
people who are 'in need of care or control' in public places and
to take them to a 'place of safety', usually a police station or a
hospital, without charging them.[6] In these circumstances, a
medical examination must take place within the next seventy-two
hours in order that arrangements may be made for remedial
treatment or appearance before a court if there is evidence that a
serious offence has been committed. In most cases, although
often rowdy and sometimes violent, the behaviour leading to
police intervention does not amount to a criminal offence, or
else constitutes a breach of the law so trivial that no prosecution
follows.[7]

Temporary confinement on these grounds is limited to the
behaviour of apparently disturbed people in public places; it
does not extend to persons whose mental state may be giving
cause for concern in private or domestic situations. Although it
is possible for a mentally disordered person, or someone whose
conduct gives the impression that he may be suffering from a

---

[5] Under the mental health legislation, patients may also be received into
guardianship in pursuance of a guardianship application. Although the courts
are able to make an order placing an offender under the guardianship of a local
health authority or of any other person, the power is seldom used in criminal
cases.

[6] N. Walker, *Sentencing: Theory, Law and Practice*, Butterworths, London,
1985, p. 323.

[7] Ibid., p. 349.

mental disorder, to be deprived of his liberty temporarily by the police rather than by the order of a court or the recommendation of a doctor, this provision is nevertheless part of the formal mental health powers of the state and is legitimized in a statute approved by Parliament.[8] If I labour this point, it is because more than a legal or procedural nicety is at issue. The dividing lines between mental abnormality and criminal conduct need to be clearly drawn and recognized if those enjoying temporal power are not tempted to use hospitals or other places of confinement as an alternative to imprisonment for the purposes of removing from circulation anyone whose continued liberty is an inconvenience or worse. English history has not generally been disfigured by the locking up of political opponents in asylums for the insane. This has not been so elsewhere; nor, unhappily, is it only the past tense that is applicable today.

The abuse of psychiatry in the Soviet Union, by which considerable numbers of political dissidents are classified as deranged, confined to mental hospitals, and subjected to compulsory treatment (including by drugs), is amply documented and has been widely condemned.[9] This odious practice has been denounced by professional medical opinion throughout the non-communist world, and as recently as the early 1980s it split the international psychiatric bodies.[10] Every national society has blemishes, but few are so repugnant as the abuse of psychiatric methods to suppress dissent. Of all the reforms that would complement the recent developments liberalizing restrictions on dissidents, none would be regarded by international professional bodies as more of a touchstone than a deliberate move away from the recourse to mental hospitals as places of confinement for those whose psychiatric condition is normal but whose

---

[8] Mental Health Act 1983, S. 136.

[9] S. Bloch and P. Reddaway, *Soviet Psychiatric Abuse*, Gollancz, London, 1984, and the same authors' *Russia's Political Hospitals*, Gollancz, London, 1977 (retitled *Psychiatric Terror*, Basic Books, New York, 1977, in the American edition); Amnesty International, *Political Abuse of Psychiatry in the USSR*, Amnesty International Index: EUR 46/01/83, 1983; and a chapter on the political use of psychiatry in J. Robitscher, *The Powers of Psychiatry*, Houghton Mifflin, Boston, 1980, pp. 319–46.

[10] Faced with the possibility of expulsion, the Soviet Union resigned from the World Psychiatric Association in 1983. See Bloch and Reddaway, *Soviet Psychiatric Abuse*, pp. 197–211.

political motives and loyalties are suspect.[11] Other cases of forcible detention of prisoners of conscience in psychiatric hospitals in Romania and Yugoslavia, as well as in the Soviet Union, were documented by Amnesty International in 1985/6. Amnesty has also reported on the way in which psychiatric knowledge has been used systematically in the disorientation and ill-treatment of political prisoners in Uruguay.[12]

II

Under the common law, there has never been an obligation that all suspected criminal offences must automatically be the subject of prosecution. The prosecution authorities have to consider whether or not there is sufficient evidence to justify proceedings and whether the public interest requires a prosecution. Nevertheless, the practice has been that the graver the offence, the less likelihood there will be that the public interest will allow of a disposal less than a prosecution, for example a caution.[13] Where there is evidence that a serious crime has been committed, medical reports may be requested at committal stage before the Magistrates' court so that the accused can be examined in advance of his or her case coming to trial. In all homicide cases, and in some other serious offences of violence to the person, it is routine for reports to be requested from two qualified psychiatrists. The starting point, therefore, is that mentally disordered persons are tried in the same courts and are subject to the same law as any other person charged with the commission of a criminal act.

Even where an accused person's state of mind is so severely disturbed as to make it impossible for him to stand trial, the question will normally be put to a jury in a Crown Court. Lack of competence to plead is a neglected aspect of the process of criminal justice. Few statistics exist beyond the knowledge that

[11] Under Stalin, psychiatric intervention was originally said to have saved some dissidents from death or the Gulags by offering diagnoses of mental illness which the psychiatrists themselves did not always believe. Unfortunately, later psychiatrists did come to believe.
[12] Amnesty International Index: AMR 52/18/83.
[13] Crown Prosecution Service, *Code for Crown Prosecutors*, HMSO, London, 1986, p. 4.

between ten and thirty persons per year have been admitted to hospital subject to restrictions over the last decade after being categorized as unfit to plead. The basis of the plea goes to the heart of the notion of criminal justice. To be guilty of a crime, an individual needs to be aware of what he or she may have done and to have had the necessary degree of intent or recklessness. Actual guilt can be established only in court and as a result of procedures which are designed to balance the interests of the offender with those of the state. Foremost among these is that an accused person should be present at his own trial, a protection against conviction and sentence *in absentia.* He should also have, to a reasonable degree, a rational understanding of the nature of the proceedings against him and the ability to consult defence counsel.[14] If the accused cannot comprehend the fact that he is on trial for a criminal offence and that he has the right to be defended, if he cannot grasp the significance of what is happening to him, then he cannot be regarded as fit to stand trial. In effect, the obligation on an accused person to attend his trial means that he should be mentally as well as physically present in court. In rare cases physical handicap may also be relevant; for instance, a person who is deaf/mute, although possessing a sufficient degree of intent to be held responsible for committing a crime, may nevertheless be unfit to plead.

Objective criteria on fitness to stand trial are hard to determine, although some work done in the United States and Canada deserves attention.[15] In one leading American case a woman was charged with a particularly heinous child murder involving torture and sexual abuse. Despite fainting or losing consciousness eleven times during the trial, and twice falling out of the witness box, the defendant was convicted, although the conviction was set aside on appeal.[16] There is merit in the proposal that the evaluation of competency should not be left

[14] This formulation follows the constitutional standard for trial competency set by the United States Supreme Court in *Dusky* v. *United States* in 1960. Quoted by R. Roesch and S. L. Golding, *Competency to Stand Trial,* University of Illinois Press, Urbana, Illinois, 1980, pp. 11 *et seq.*

[15] Roesch and Golding, *Competency to Stand Trial,* is the leading work. In 1984 the same authors submitted a detailed report for the National Institute of Mental Health on 'Evaluation of Procedures for Assessing Competency to Stand Trial'.

[16] *People* v. *Berling,* 1953. Cited in Roesch and Golding, *Competency to Stand Trial,* p. 19.

solely to clinical assessment by doctors, and that a defendant's cognitive, behavioural, and affective capacities should be evaluated by lawyers and social workers as well as by doctors in relation to the actual charges. But it is admitted that in most US jurisdictions judges rarely base their decisions on anything more than the concluding statement in the psychiatric report before them, and that the proceedings seldom occupy much of the time of the court.

These cases apart, once an accused person reaches the court, any query concerning his mental capacity will relate in the large majority of cases to the question of intent or *mens rea*. Since it is a precept of English law that no one can be punished for a crime he did not consciously intend to commit, or which was committed without due foresight of the consequences,[17] criminal responsibility must be proved beyond reasonable doubt. It is at this point that an awkward issue arises. Just as it is evident that there are few clear-cut dividing lines between sanity and insanity, but rather a range of gradations between two poles, so an individual's mental capacity will vary. Sometimes he will be closer to the end of what a court might regard as a state of responsibility for his own actions; sometimes he may veer away towards the extremity of irresponsibility. Who can say with any confidence what another person, probably unknown to him previously, did or did not intend to do, possibly at a moment of emotional crisis or where the state of his mind was disturbed, and often in a situation where no independent witness was present? Thus, objective evidence will be sought from expert medical witnesses as to what may be known about the defendant's mental state before and after the crime. Is there anything on record in the past, either previous offences or psychiatric treatment, or in his behaviour before or immediately after the crime, which may help to explain what took place? If so, the court will want to hear it, especially in guilty pleas where there is no risk of prejudicing a defendant's trial.

In 1959 an untidy jumble of nineteenth-century legislation was

---

[17] Recklessness, like intention, requires foresight of certain consequences as being inevitable, or probable, or sometimes merely possible. Unlike intention, however, recklessness involves no desire that these consequences should result or will to bring them about. D. M. Walker, *The Oxford Companion to Law*, Clarendon Press, Oxford, 1980, p. 1042.

repealed and replaced with a comprehensive Mental Health Act, part of which made provision for mentally disordered persons convicted of criminal offences. Section 60 listed four separate categories of mental disorder, namely: mental illness, psychopathic disorder, subnormality, and severe subnormality. With one change, the same phraseology was retained and re-enacted in Section 37 of the Mental Health Act 1983; the terms 'mental impairment' and 'severe mental impairment' have been substituted for 'subnormality' and 'severe subnormality'. If, after conviction and having been diagnosed by two registered medical practitioners (one of whom must have been approved as having special experience in the diagnosis or treatment of mental disorder), the court is satisfied that the offender falls into one of the categories mentioned above, it may make a hospital order rather than sending him to prison or imposing a non-custodial penalty. Such an order means that an offender will be detained in hospital, where he will receive treatment for his condition, for as long as is necessary. Hospital orders can be made only where the court is satisfied that firm arrangements exist for the admission of an offender to hospital within twenty-eight days, and this is often a limiting factor. At what is judged to be the appropriate stage in his treatment, he can be discharged by the responsible medical officer or by a Mental Health Review Tribunal. The degree of security will vary according to the nature of his mental disorder and the gravity of the offence.

In the graver cases, the higher courts have an additional power to make a restriction order 'for the protection of the public from serious harm', so that the patient is compulsorily detained in secure hospital accommodation for either a specified or an indefinite period of time ('without limit of time'). The reference to 'serious harm' in the Mental Health Act 1983 resulted from a recommendation by the Butler Committee, which concluded that restriction orders under the 1959 Act were sometimes made when they were not strictly necessary. Revised wording was proposed to make it clear that the purpose of the restriction order—a severe control—was to protect the public from serious harm.[18] In making such an order, the court will have regard to the

---

[18] *Report of the Committee on Mentally Abnormal Offenders*, p. 198.

nature of the offence, the antecedents of the offender, and the risk of his committing further offences if set at large.

Under the 1959 Act, the consent of the Home Secretary was required before any restricted patient could be discharged or moved to another hospital. An important change in the 1983 Act, following a judgment in the European Court of Human Rights,[19] was that the Home Secretary no longer had the sole power to release restricted patients. If a Mental Health Review Tribunal is satisified that a restricted patient is not suffering from a mental disorder of a form that makes it appropriate for him to be detained in hospital for medical treatment, it is obliged to direct his discharge.[20] The order may be absolute or conditional, and many offender-patients remain subject to supervision and recall after discharge. We shall see how this can work in practice later in this chapter. It should be noted that hospital orders with or without restrictions may be invoked only where the offence is one that carries a sentence of imprisonment to be decided by the court.[21] Therefore the liability of an individual to lose his liberty, if found guilty, will not necessarily change irrespective of whether he is dealt with under the penal powers or the mental health powers of the state. It is the place of confinement and the treatment to which he will be subject that will differ, as will the procedures for his ultimate release or discharge.

From this short account, it will be seen that the handling of disordered offenders straddles the institutions of criminal justice and medical care. Difficult issues abound: practical as well as principled. For example, what was in many ways an enlightened attempt to disentangle the sick from the wicked, and to divert them into appropriate streams for treatment or punishment, coincided with a deliberate move in the health service towards the treatment of as many mentally disordered people as possible

[19] In the case of *X* v. *The United Kingdom* in October 1982, the European Court of Human Rights ruled that a restricted patient was entitled to have the lawfulness of his compulsory confinement in a psychiatric hospital reviewed either by a court or by another body of a judicial character.

[20] The operation of the Mental Health Review Tribunals has been analysed by Dr J. Peay in a paper entitled 'Psychiatry and the Law—Who Controls Who?' presented at the Second International Congress on Psychiatry, Law, and Ethics in Tel Aviv in 1986.

[21] Criminal offences for which the penalty is prescribed by law, notably murder and treason, are excluded.

in the community rather than in hospital.[22] Patients requiring treatment in hospital, it was felt, should be admitted voluntarily and discharged once treated. Ambitious and honourable as this policy was, it nevertheless presented immediate difficulties for offenders diagnosed as suffering from one of the four categories of mental disorder recognized in the Mental Health Act 1959. The courts could make a probation order with a condition of psychiatric treatment, a disposal that was increasingly used for non-dangerous offenders; but in the more serious cases the risk to the public of further offending might be regarded as being too great to permit a potentially dangerous offender to remain at liberty, even if supervised. Therefore a degree of security would often be sought by the courts in determining how to dispose of the numerous mentally disturbed offenders appearing before them. But at this very moment, up and down the country, hospitals were being encouraged to throw open their doors and unlock their psychiatric wards. Patients requiring secure conditions were not easy to accommodate in national health service (NHS) hospitals. Their presence disrupted the atmosphere and made therapy more difficult. There was also growing opposition from staff unions. These were the underlying causes of the ever-widening chasm which opened up between the aims and priorities of the health service and those of the penal system.

Some idea of the scope of offending can be obtained from a study of annual hospital admissions subject to restrictions on release, together with hospital orders imposed by the courts for indictable or non-indictable offences each year. Table 1 shows the position as it was in 1984, a broad picture of some 1,100 persons suffering from mental disorders of varying kinds being admitted to hospital for treatment with or without restrictions.

[22] The *Annual Report of the Department of Health and Social Security for 1975* referred to proposals for 'a pattern of services which would involve the gradual phasing out of the old and large mental hospitals, and their replacement by services based closer to the patient's home, and better adapted to his individual needs. Hospital treatment would be based on a psychiatric unit, including a day hospital, forming part of the general hospital for the district, but there would also be the need for staffed and unstaffed homes and hostels in the locality, as part of local authority social services, to provide long- and short-term support. Greater emphasis would also be given to helping those who are or have been mentally ill to live in their own homes.' Cmnd. 6565, HMSO, London, 1976, p. 39.

TABLE 1. *Mentally disordered offenders: hospital admissions, 1984*

	Subject to restrictions	Without restrictions
Indictable	272	641
Non-indictable	11	178
Civil prisoners	2	0
TOTAL	285	819

By no means all of the total of 285 who were subject to restrictions reached hospital as a result of a court order, since 129 were transferred from Prison Department establishments (46 before sentence and 83 after). Of all admissions in this category, 80% were regarded as suffering from mental illness, 14% from psychopathic disorder, and about 5% from mental impairment or severe impairment.

A word of caution is necessary about the interpretation of these statistics. As a numerical indicator of how many offenders are disordered, they are certainly a substantial underestimate, since the courts may be constrained from making hospital orders where they would otherwise do so because of the unavailability of beds, while many disordered offenders make their way through the penal system without having their disorder either identified or diagnosed. This is particularly likely to apply in the case of non-violent property offenders who rarely attract psychiatric attention. There is, in fact, some tentative evidence to suggest that up to about one-third of the prison population of sentenced prisoners may be regarded as psychiatric cases who need and are willing to accept treatment for their mental condition. As many as one in five prisoners in a Home Office sample had received previous psychiatric treatment in the NHS.[23]

An analysis of the offences committed by mentally disordered offenders admitted to hospital in 1984, subject to restrictions on release, reveals the pattern shown in Table 2. More than half of

[23] J. Gunn, G. Robertson, S. Dell, and C. Way, *Psychiatric Aspects of Imprisonment*, Academic Press, London, 1978, pp. 209–32.

TABLE 2. *Mentally disordered offenders: hospital admissions subject to restrictions, by offence group, 1984*

Offence	No.	%
Violence against the person:		
Homicide	58	20
Other	99	35
Sexual offences	21	7
Burglary	22	8
Robbery	7	2
Theft, handling stolen goods	13	5
Fraud, forgery	2	1
Criminal damage:		
Arson	43	15
Other	3	1
Other indictable	4	1
Summary offences	11	4

all patients had been convicted of violent offences. This rose to 70% for the older patients (over forty), and fell to less than a quarter for the under-twenty-ones, of whom one-third had been convicted of an offence of arson.

The offences committed by the total of 819 persons subject to hospital orders without restrictions on release are shown in Table 3. Once again, violence against the person still comprises the largest single group of offences, but it is far less common than among those subject to restrictions on release. Homicide is rare, and less serious offences such as theft and summary offences predominate.

The shortage of hospital places for mentally disordered offenders who require in-patient treatment has been a running sore between the courts and the medical profession. Over and over again, a sentencing judge or bench of magistrates will bemoan the lack of adequate hospital facilities where a disordered offender requires a degree of confinement and a form of treatment that he is not likely to get in prison. Apart from the four special hospitals—Broadmoor, Rampton, Moss Side, and Park Lane, which are managed directly by the Department of Health and Social Security (DHSS) and are reserved for the highest risk cases—doctors with mentally disturbed patients

TABLE 3. *Mentally disordered offenders: hospital admissions without restrictions, by offence group, 1984*

Offence	No.	%
Violence against the person:		
Homicide	13	2
Other	240	29
Sexual offences	31	4
Burglary	52	6
Robbery	4	1
Theft, handling stolen goods	151	18
Fraud, forgery	34	4
Criminal damage:		
Arson	52	6
Other	43	5
Other indictable	21	3
Summary offences	178	22

facing court proceedings must shop around for a place. Often they have been as frustrated as the court in their inability to arrange for appropriate hospital admissions. Since the Mental Health Act 1983, the court may hear evidence from the regional health authority as to why a place is not available. Lack of resources, lack of security, and nursing staff or union opposition to the admission of a larger number of potentially disruptive (if not actually dangerous) patients than they feel they can cope with are the reasons most commonly advanced. Moreover, the capacity of the special hospitals is reduced by a substantial number of non-offender patients who have been transferred from conventional psychiatric hospitals.

In April 1974, in an attempt to ease the growing problem and also as a way of providing a staging post between the special hospitals and the open regime of local psychiatric hospitals, an interim recommendation was made by the Butler Committee for the establishment, as a matter of urgency, of secure hospital units in the area of each regional health authority.[24] An internal DHSS

[24] *Interim Report of the Committee on Mentally Abnormal Offenders* (Cmnd. 5698), HMSO, London, 1974, p. 4.

working party had reached the same conclusion, and with grati-
fyingly little delay the Government accepted the proposal in July
of the same year. A shrewd old hand had not lost its cunning in
influencing public policy. Additional funds were promised by
central government to cover the capital costs of building and
equipping the new units, although the regional health authorities
were originally expected to meet the running costs from revenue.
In order not to delay the provision of secure units until purpose-
built units were available, a departmental circular was issued
emphasizing the need for urgent action and requesting that
suitable interim arrangements should be made for treating
patients in conditions of security.

Despite a favourable reception by informed, non-specialist
opinion, and the endorsement of both major parties, the idea of
secure units for offenders being placed in or adjacent to NHS
hospitals operating the open door philosophy went against the
grain in the health service. Many regarded it as a retrograde step,
some even going so far as to speak of the Government's policies
of 'bribing' reluctant hospitals to set up segregated units run by
forensic psychiatrists as being wrong in principle and unlikely to
fill the gap in caring for awkward and embarrassing patients.[25]
A year passed with little progress. In its final report in October
1975, the Butler Committee tartly observed: 'What causes us
concern is that in some parts of the country there have been
indications of unwillingness either to formulate plans at all, or to
give our proposals the priority which we thought and the
Government agreed to be necessary.'[26]

Although progress was painfully slow, bit by bit the
programme inched forward, with more money being found to
contribute towards running costs, while health service and staff
resistance was gradually worn down. The first interim secure
units were set up in refurbished parts of existing hospitals. Once
sites could be found and the fears of local populations overcome,
new purpose-built units were constructed and opened. By early
1986, more than ten years after the recommendations were
accepted by the Government calling for urgent implementation,
permanent secure units were still not functioning in every regional

[25] P. D. Scott, *Has Psychiatry Failed?*, Institute for the Study and Treatment
of Delinquency, London, 1975, p. 11.
[26] *Report of the Committee on Mentally Abnormal Offenders*, p. 3.

health authority.[27] With a likely bed complement by the end of 1986/7 of 633 beds for patients requiring intensive treatment and rehabilitation in secure conditions, whether or not they have been convicted of criminal offences, as against Butler's target of 2,000, and with part of the additional revenue funding diverted into general psychiatric services, the rearguard action is not yet spent. The issue was one on which Lord Butler, with a life-long interest in penal reform, felt deeply. One of his last appearances in Parliament was when he came to the House of Lords, at some personal inconvenience, to speak in support of a question critical of the lack of progress being made.

The structure of government departments, and the division of responsibilities between them, has done nothing to alleviate the delays or solve the problems. The Home Office is concerned only with convicted offenders, some of whom, if considered likely to benefit from treatment, have been routed by the courts into the hospital system. The treatment they receive is provided by the national health service under the policy direction of the Department of Health and Social Security. The Home Office wishes to ensure that the conditions in which restricted patients who have been found guilty of a criminal offence are treated are adequately secure. Some of these people are violent and dangerous, and the safety of the public, as well as the impact on public opinion in the event of a restricted patient absconding, is uppermost in its outlook. Other restricted patients may require lesser security, some being accommodated in NHS hospitals with lighter restraints.

In contrast, the Department of Health looks on patients primarily as sick people in need of remedial treatment. Medical and nursing staffs at the special hospitals and secure units are well aware of the uniqueness of their situation and have evolved forms of assessment and treatment aimed at meeting the requirements of their disturbed or mentally ill charges, irrespective of whether or not they have been through the courts. But to DHSS administrators at headquarters, responsible for decisions on resources and policy, the presence of offenders in the hospital system is a minor irritation, an annoying interruption of their

[27] Details of regional secure units in operation or near completion were published in a Written Answer to a Question in the House of Lords in March 1986. *Parl. Debates*, HL, 472 (5th ser.), cols. 295–6, 5 Mar. 1986.

main activities. Compared with the total of NHS patients, only a small number of people are involved. The Department knows little about the courts and the prisons and is not very interested anyway; there are other more pressing matters claiming attention. It is tiresome and time-consuming to be diverted from the major issues of health care by the very difficult, but relatively small-scale, problems of treating mentally disordered offenders who are subject to court orders.

One by-product of the move towards regional secure units, slow as it has been, is the growing number of well qualified psychiatrists with a particular interest, and developing expertise, in forensic problems. This trend is influencing the situation in a more favourable direction, not only as a result of their work at the special hospitals and regional secure units, but in providing clinical opinions and advice to their colleagues throughout the national health service.

III

The way in which mentally disordered offenders should be treated has always been a difficult area where the territorial prerogatives of medicine and the law overlap. They do so, moreover, not simply in instances that can be resolved by dialogue and (where this can be achieved) consensus between professionals, but under an ominous cloud of public opinion. Periodically, the cloud bursts and a storm of popular reaction erupts. Three notorious incidents, involving attempts on the lives of a British prime minister, a reigning sovereign, and a president of the United States, each led to changes in the law relating to insanity. The first occurred in 1843, when Daniel M'Naghten was sent to the Bethlem Hospital—the original Bedlam[28]—after shooting and killing Sir Robert Peel's private secretary on the mistaken assumption that he was the Prime Minister. M'Naghten

---

[28] Founded as early as 1375, by the seventeenth and eighteenth centuries the Bethlem Hospital had become the principal repository for the mentally afflicted in London and was accepting certain patients by order of the Government. A block for the containment of criminal lunatics was built in the early part of the nineteenth century, and this function continued until the opening of Broadmoor. See Scott, *Has Psychiatry Failed?*, p. 11.

remained at Bethlem for virtually the rest of his life, being transferred to the newly opened Broadmoor Hospital a year before his death from tuberculosis in 1865. Whether he was a deranged and irrational assassin or, as has been argued more recently, an extreme radical who was politically motivated to kill Peel as a protest against Tory policies,[29] M'Naghten's case made legal history. At his trial, once the medical evidence had been given, he was acquitted of murder on the grounds of his insanity by a jury that did not even find it necessary to leave the court room at the Old Bailey.

The verdict caused a political furore. Public opinion was outraged that M'Naghten had been found not guilty and had escaped the gallows. From now on, it was proclaimed, no statesman or politician could safely walk the streets for fear of assassination. Parliament was alarmed and the young Queen dismayed. Like many of her subjects, she found it hard to accept the 'not guilty' verdict, a belief that remained unchanged and undiminished in intensity until the same issue recurred, dramatically and personally for Queen Victoria, forty years later in her reign. In the aftermath of the M'Naghten verdict, the House of Lords summoned all the judges of the Supreme Court of Judicature to appear before the House in person and put to them five questions designed to elicit and clarify the law on insanity. The answers they received, known as the 'M'Naghten rules', have remained the test of insanity which the courts have applied ever since. The rules prescribed that under English law every man is presumed to be sane and responsible for his criminal act unless the contrary is proved. To establish a defence of insanity, it must be clearly proved that at the time of committing the act the accused person was labouring under such defect of reason from disease of the mind as not to know the nature and the quality of the act, or, if he did know it, not to know that what he was doing

---

[29] In a book titled *Knowing Right from Wrong: The Insanity Defence of Daniel McNaughtan* (Free Press, New York/Collier Macmillan, London, 1981), Richard Moran argues that McNaughtan (the spelling he prefers, and which is accepted by Professor Nigel Walker) was a young Scottish radical who supported the Chartists and was obsessed by hatred of the Tories. He had planned to murder Peel as a political protest, but shot the wrong man as a result of mistaken identity.

was wrong.[30] Later judgments in the courts established that 'wrong' in this context meant not morally wrong (about which opinions differ), but punishable by law.

In 1882, as Queen Victoria was about to descend from her carriage at Windsor Station, a man named Maclean fired a loaded pistol at her. Fortunately, she was unharmed while her assailant was attacked by two Eton boys with their umbrellas. It was not the first time the Queen had been shot at or assaulted by her deranged subjects. On five previous occasions pistols had been fired at her, and in a particularly scandalous incident earlier in her reign she had been struck on the head with a brass-topped cane by an ex-cavalry officer in Piccadilly.[31] Now her patience was finally exhausted, and she insisted on a change in the law so that acquittals on grounds of insanity would be replaced by verdicts of guilty but insane. In *Crime and Insanity in England,* Nigel Walker records that at first Gladstone was cautious, saying that, being ignorant of the law, he hesitated to venture an opinion (just the sort of reply that infuriated the Queen); but before long he too became convinced that a change in the form of verdict was justified. The Prime Minister's letter to the Queen informing Her Majesty of the introduction of a Bill to effect 'this most proper change' in the law of criminal lunacy is reproduced in facsimile in the first volume of Professor Walker's splendid history.[32] A few months later the Trial of Lunatics Act 1883 was on the statute book, Opposition as well as Government spokesmen in Parliament having being privately apprised of its royal provenance.

The change from acquittals on grounds of insanity to verdicts of guilty but insane (which endured until 1964[33]) was as much a response to events, and the resulting public clamour, as a rational development in the law. In this it was mirrored exactly,

[30] N. Walker, *Crime and Insanity in England.* Vol. 1, *The Historical Perspective* (University Press, Edinburgh, 1968) contains an account of M'Naghten's case and the resulting rules at pp. 84–103.

[31] Ibid., pp. 187–8.

[32] Ibid., p. 191. Nigel Walker succeeded Sir Leon Radzinowicz as Wolfson Professor of Criminology at Cambridge, 1973–84, and was Director of the Institute of Criminology, 1973–80.

[33] The Criminal Procedure (Insanity) Act 1964 restored the earlier formula of not guilty on ground of insanity following a recommendation by the Criminal Law Revision Committee.

almost a century later, in an attempt on the life of President Reagan. Only ten weeks after his inauguration in 1981, the President was shot and wounded as he left a Washington hotel after addressing a trade union convention. Three men accompanying him were also hit with bullets fired from a .22 calibre revolver, one of them, the President's press secretary, being shot in the head. Mr Reagan was driven immediately to hospital where in the course of a two-hour operation a bullet was removed from his left lung. He made a good recovery and shortly afterwards was able to resume his presidential functions. Unlike the fatal attack on his predecessor, John Kennedy, whose assassin never lived to stand trial,[34] there was no difficulty in identifying the assailant, who was standing among a group of reporters at the hotel exit. John Hinckley, aged twenty-five, was immediately arrested, and after being charged with the attempted assassination of the President was remanded by a federal district court to an institution for mental examination to determine his sanity. Over a year later, after an eight-week trial estimated to have cost in excess of $3 million, in which teams of psychiatric experts vigorously contradicted each other's diagnoses, Hinckley was found not guilty because of insanity and committed indefinitely to a federal mental hospital.

Once again, as in nineteenth-century Britain, the verdict was greeted by the public with a mixture of incredulity and outrage; once again, the prime target expressed his own concern over the law on insanity; and once again, the verdict prompted attempts to revise or abolish the insanity defence. The Insanity Defense Reform Act 1984 was the result. Never previously the subject of legislation enacted by Congress, the defence of insanity had rested on case law following closely the M'Naghten rules. Although insanity still remains as a defence to prosecution under any federal statute, it is more tightly defined, with the burden of proof on the defendant, who now needs to demonstrate his insanity by 'clear and convincing evidence'.[35]

[34] Assuming, as did the Commission presided over by Chief Justice Earl Warren, that the assassin of President Kennedy was Lee Harvey Oswald.

[35] Insanity Defense Reform Act 1984 (Pub. L. 98–473, Title II, S. 401, 12 Oct. 1984, 98 Stat. 2057). For a full description of the insanity defence in US criminal law see A. S. Goldstein, 'Excuse: Insanity' in S. H. Kadish (ed.), *Encyclopedia of Crime and Justice*, Vol. 2, Free Press, New York, 1983, pp. 735–41.

IV

Fascinating as these historical parallels are, they are barely relevant to current practice in the English courts. Since the introduction of diminished responsibility in 1957, followed by the suspension of capital punishment in 1965 (and its abolition in 1969), the insanity defence has fallen into disuse. When the alternative in homicide cases was the death penalty, an indeterminate period of confinement in a mental hospital had obvious attractions for the criminally insane.

At this point in the narrative it should be noted that well embedded in the common law tradition was the notion that, whatever the circumstances, no accused person found to be criminally insane should be executed. The mob might be disappointed, but neither Church nor State would tolerate it. According to an American scholar, Norval Morris, 'As a matter of legal theory, the insane were denied the release of capital punishment ... A jurisprudence sensitive to theological concerns allowed for repentance on the gallows or before execution, and the insane were in no position to repent.'[36] So long as capital punishment existed, the insanity defence was of importance in avoiding the gallows, and consequently it provoked much public interest and controversy. But even in the United States, where the death penalty remains legal in thirty-eight states (although not in the federal jurisdiction under which Hinckley was tried), the insanity defence is now rarely asserted.

When abolition came in Britain, the balance of advantage changed. No longer was disposal under the mental health powers of the state preferable to penal sanctions, since the worst that could happen to a mentally disordered offender who preferred to take his chances in the penal system was that he might receive a life sentence. In practice, this was equivalent to detention for an unlimited period of time in a special hospital. In both cases, the Home Secretary's consent would be required for release. From the standpoint of loss of liberty, there was nothing in it. A conviction of murder would result in a mandatory life sentence, a provision that is discussed—and criticized—in the next chapter.

[36] N. Morris, *Madness and the Criminal Law*, University of Chicago Press, Chicago and London, 1982, p. 130. Norval Morris is Julius Kreeger Professor of Law and Criminology at the University of Chicago.

But there was always a chance that the charge might be reduced to manslaughter, in which case the court had discretion on conviction to impose any sentence ranging from life imprisonment to a probation order. Moreover, a new and powerful weapon was at hand: the defence of diminished responsibility.

Besides making a distinction between categories of capital and non-capital murder, an approach that was soon discredited by experience, the Homicide Act 1957 introduced a new defence borrowed from Scotland. For the best part of a century, as an expedient rather than a conscious addition to Scottish legal doctrine, there had been a third alternative open to the courts north of the border in deciding how to deal with a mentally disordered offender. English law limited the courts to answering a straightforward question (provided the accused person was fit to stand trial at all): either he was sane and responsible for his acts or omissions, or he was not.[37] Faced with this stark choice, the Scots had added a third possibility falling midway between the other two: that an accused person might have been sufficiently sane to have known what he was doing when he committed the crime, and to have been aware that it was wrong, but that his mental state at the time was such that the degree of punishment should be reduced. Thus, an accused person's state of mind might be accepted as an extenuating circumstance—not enough to warrant his acquittal on grounds of insanity, but sufficient to reduce the charge from murder to the offence known to Scots law as 'culpable homicide'.[38] The ingeniousness of this device was that it was not the death penalty that was mitigated, but that another and lesser charge was substituted for murder. The theoretical basis might be questioned, but judges were provided with an escape route leading away from the gallows in compassionate cases. The idea was not peculiar to Scots law. Elsewhere in Europe, from the time of Matthaeus onwards, institutional writers had argued that there are states of mind in which people break the law without being so disordered as to be blameless, or so in control of themselves as to be fully culpable. The mental state of mothers who kill their infants while depressed shortly after birth is one familiar example, and the defence of provocation is another.

[37] Walker, *Crime and Insanity in England*, Vol. 1, p. 147.
[38] Ibid., pp. 138–46.

This was the precedent that was imported into English law in the compromise Bill which became the Homicide Act 1957. Had Parliament abolished capital punishment outright, rather than preserving it for certain types of murder, it is unlikely that the defence of diminished responsibility would have been introduced at all. What was seen by some at the time as little more than a makeweight in highly controversial legislation was to become a cornerstone of the penal system in practice. And so the statute book is made up. The convoluted wording of Section 2 of the Homicide Act runs as follows:

> where a person kills or is party to the killing of another, he shall not be convicted of murder if he was suffering from such abnormality of mind (whether arising from a condition of arrested or retarded development of mind or any inherent causes or induced by disease or injury) as substantially impaired his mental responsibility for his acts ...

It is for the defence to raise the issue, and if it is successful in doing so the accused person becomes liable to be convicted of manslaughter instead of murder. Initially the courts insisted that the decision as to whether the charge should be murder or manslaughter should be left to the jury having heard the medical evidence for the defence, which might be challenged by the prosecution.[39] But within five years, a large majority of pleas of not guilty to murder by reason of diminished responsibility but guilty to manslaughter were accepted by the Crown. This remains the situation today, although periodically a notorious trial occurs, where the extent and full horror of the killings is quite out of the ordinary, as were the cases of Peter Sutcliffe (the Yorkshire Ripper) at the Central Criminal Court in 1981, and Dennis Nilsen in 1983. In such circumstances, the judge may rule that the question of whether or not the accused person's abnormality of mind is such as to impair his mental responsibility for his acts should be put to the jury. In each of these trials multiple killings were involved, public interest was intense, and murder verdicts were returned.[40]

Although the concept of diminished responsibility is not easily

[39] Ibid., pp. 150–4.
[40] B. Masters, *Killing for Company: The Case of Dennis Nilsen*, Jonathan Cape, London, 1985. Sutcliffe was transferred from prison to a special hospital in 1984. Nilsen has remained in prison, and in December 1986 there was no plan to transfer him to a special hospital.

reconcilable with more clear-cut interpretations of the doctrine of criminal responsibility, nevertheless, it has provided a relatively safe path between extremes for mentally disordered persons standing trial for taking the lives of others. The defence applies only in homicide cases, and since the suspension of capital punishment for all forms of murder in 1965 its continued use depends on the mandatory life sentence for murder. Following conviction on any other serious charge of violence, the court not only can but must take into account all relevant factors, including (where appropriate) a convicted offender's mental state,[41] before determining what the sentence should be. Sometimes the court may decide on a hospital order; more often a criminal sanction will be imposed which may or may not result in loss of liberty. If the separate offences of murder and manslaughter were to be combined in a single offence of unjustifiable homicide, with the penalty at the discretion of the court within any limits laid down by Parliament, then the justification for the special defence of diminished responsibility would disappear. But until that day, and as a prudent safeguard against the possible reintroduction of capital punishment, it should remain. It is certainly well used, as Table 4 demonstrates. The table summarizes, over the five years 1981–5, the number of men and women originally charged with murder, but found guilty of manslaughter by virtue of diminished responsibility. Sub-totals of male and female offenders and the type of sentence given by the courts (including hospital orders) are shown, as are comparative statistics, in Table 5, of the larger number of manslaughter convictions, excluding diminished responsibility, over the same period.

The fact that less than half of all Section 2 manslaughter convictions lead to hospital orders (see Table 4) may derive less from judicial preferences for penal sanctions than from medical opinion as to suitability for remedial treatment and the availability of beds. In the early years of the diminished responsibility defence, around two-thirds of those convicted of manslaughter on these grounds were hospitalized with restrictions on

---

[41] In 1980 there were approximately 8,000 custodial remands for psychiatric examination, plus an unknown number of remands on bail. Less than one-quarter led to psychiatric disposals by the courts. Walker, *Sentencing: Theory, Law, and Practice*, p. 33.

TABLE 4. *Manslaughter convictions: diminished responsibility, England and Wales, 1981–5*

	1981	1982	1983	1984	1985
Manslaughter convictions (diminished responsibility) *of which*:	84	90	84	72	65
Male	64	76	68	57	49
Female	20	14	16	15	16
Sentences given:					
Life imprisonment	14	24	4	5	3
Over 10 years	—	—	—	—	—
5–10 years	7	7	7	5	7
Up to 5 years	13	9	18	16	10
Youth custody/Borstal	—	—	1	1	—
Hospital orders *of which*:	38	28	37	35	31
Hospital orders with restrictions (Sections 37/41 MHA 1983)[a]	26	23	29	25	21
Hospital orders (Section 37 MHA 1983)[a]	12	5	8	10	10
Detained under Section 53 CYPA 1933[b]	—	—	1	—	1
Other penalties	12	22	16	10	13

[a] Mental Health Act 1983.
[b] Children and Young Persons Act 1933.

discharge, while one-third received sentences of imprisonment. The high point was reached in the two years 1968 and 1969, when as many as 70% of diminished responsibility manslaughter convictions resulted in hospital orders. The percentage then fell steadily throughout the 1970s: from 55% in 1970–3, to 35% in 1974–7, and levelling off at about 25% by the end of the decade.[42] In the early 1980s it began to rise again, reaching just over one-third in 1983 and 1984, and just under one-third in

[42] S. Dell, *Murder into Manslaughter* (Maudsley Monographs no. 27), University Press, Oxford, 1984; reviewed in *British Journal of Criminology*, 25 (1985), 404–5.

TABLE 5. *Manslaughter convictions: excluding diminished responsibility, England and Wales, 1981–5*

	1981	1982	1983	1984	1985
Manslaughter convictions (excluding diminished responsibility) *of which*:	167	182	164	168	200
Male	146	158	150	142	175
Female	21	24	14	26	25
Sentences given:					
Life imprisonment	3	2	1	2	2
Over 10 years	1	6	6	5	6
5–10 years	26	29	31	23	34
Up to 5 years	103	97	74	78	88
Youth custody/Borstal	2	2	19	33	37
Hospital orders *of which*:	3	4	—	1	6
Hospital orders with restrictions (Sections 37/41 MHA 1983)[a]	1	4	—	—	4
Hospital order (Section 37 MHA 1983)[a]	2	—	—	1	2
Detained under Section 53 CYPA 1933[b]	—	—	5	3	6
Other penalties	29	42	28	23	21

[a] Mental Health Act 1983.
[b] Children and Young Persons Act 1933.

1985.[43] Thus the position today is that between one-quarter and one-third of those whose plea of diminished responsibility is accepted by the court are given hospital orders with restrictions on discharge under Sections 37–41 of the Mental Health Act 1983, while a smaller number receive hospital orders without restrictions. Since the overall numbers convicted of man-slaughter on grounds of diminished responsibility have not fluctuated to any marked extent, and since the mental disorders

[43] *Criminal Statistics (England and Wales) 1985* (Cm. 10), HMSO, London, 1986. The statistics contained in Table 4.8 on p. 65 have been compiled on a different basis from those contained in Tables 4 and 5 in this chapter.

from which they suffer (with mental illness, notably schizo-
phrenia and depression, heading the list) do not seem to be on
the increase, all the indications are that this shift towards the
penal system is a result of the changing outlook in the health
service, rather than as a consequence of sentencing policies
based on punitive considerations. The increasing emphasis
placed on treatability, especially in cases of psychopathic
disorder, is also likely to have been a relevant factor.

Whatever its causes, the bias towards punishment is to some
extent balanced by the fact that the number of diminished
responsibility offenders who have been sentenced to life
imprisonment was much lower in 1983–5, five or fewer in each
year, compared with an average of twenty a year during 1975–
82.[44] It is also true that there is no unanimity of opinion as to the
greater suitability of hospitals as compared with prisons for the
treatment and containment of certain types of mentally disordered offenders, especially those who are habitually violent.
An authority as well regarded as the late Dr Peter Scott[45] used to
aver that the proper place for such offenders was in a prison
hospital, and that it was preferable to make prisons more like
hospitals for this purpose than to make hospitals more like
prisons.[46]

V

At each stage in the confinement, treatment, and finally
discharge of a mentally disordered offender, issues of principle
abound. There are few yardsticks beyond the universal requirement that the relevant statutory provisions must be observed and
orders of the courts enforced. But, as we have seen, even these
apparently fixed points in the penal firmament are less definite
than might be expected. If the very meaning of psychopathic (or
personality) disorder is contested, as it has been from the time of

[44] Ibid., p. 57.
[45] P. D. Scott, FRCP, FRCPsych. (1914–77), Consultant Physician, Bethlem
Royal and Maudsley Hospitals, was Consultant Forensic Psychiatrist to the
Home Office and a member of the Advisory Council on the Penal System.
[46] R. Mark and P. D. Scott, *The Disease of Crime: Punishment or Treatment?*
Royal Society of Medicine, London, 1972, pp. 36 *et seq.*

Hippocrates and Plato,[47] it is hardly surprising that some psychopaths end up in prison while others, whose condition may not be much different, go to hospital for treatment. As it is used today, 'psychopathic disorder' is more of a generic term for the purposes of legal classification than a specific diagnosis. It includes patients suffering from serious pathology of the mind of many different varieties, whose only common characteristic is their markedly anti-social behaviour.[48] Moreover, each year a substantial but incalculable number of offenders with mental disorders escapes recognition and diagnosis altogether.

When decisions are taken which are of vital importance to the lives of inmates in prison or patients in hospital, the exercise of discretion and personal judgment jostles with professional opinions and practical requirements. The desire to do justice and show compassion to an individual in his special situation may conflict with the imperatives of fairness and equality in considering one man's case in a way that is consistent with others. Although the ethos of separate institutions will vary greatly, almost all of those who have the task of supervising or treating mentally disordered offenders want to fulfil their tasks decently and capably. Given the adverse conditions in which they often work, the woefully inadequate facilities at their disposal, and the pressures and provocation they may endure, it is inevitable that some lapses will occur.

As the report of the review set up following allegations of ill-treatment of patients at Rampton Hospital showed, it can be failures in the regime rather than lack of individual care and attention that are the root cause of some of the most ineradicable problems in the treatment of a class of patient for whom society asks little more than they be safely locked up. The report, published in 1980, referred to the perils of institutional inertia, accentuated by size and remoteness. The geographical remoteness (seen as an advantage when the hospital was opened in 1912 as the first criminal lunatic asylum in the North of England), and a nationwide catchment area, contributed to an organizational and professional isolation. The chairman of the

[47] *Report of the Committee on Mentally Abnormal Offenders* included a lengthy historical review of psychopathic disorder at pp. 77–100.
[48] See letter to the Editor from Dr J. R. Hamilton, Medical Director, Broadmoor Hospital, *The Times*, 14 July 1986.

review team commented in his covering letter to the Secretary of State for Social Services: 'there is no doubt that changes need to be made at Rampton. The hospital appears to have been a backwater and the main currents of thought about the care of mental patients have passed it by'.[49] From first-hand experience, I know what progress has been made since those words were written. The patient population has come down from 819 on 30 June 1980, with an official bed complement of 1,051, to about 550 in June 1986, with an official bed complement of 706. Over the same period, the nursing establishment was increased from 666 to 722. The combination of these two factors resulted in a greatly improved nursing staff/patient ratio. There were also increases in the establishment of medical staff, social workers, and psychologists. Money has been spent on buildings and facilities to modernize and improve the physical environment, and a Standing Review Board has been set up to oversee the implementation of the review team's recommendations and to assume management functions on the spot which had been exercised previously by DHSS officials in London.

Despite these changes, Rampton remains a large and isolated institution where custodial attitudes of the past linger on, particularly among the nursing staff, many of whom have given their working lives to the hospital.[50] But there is now an effective management team firmly dedicated to the principles of therapeutic treatment in a secure setting. Painful though the publicity must have been for Rampton staff to bear, and unfair as many of the criticisms broadcast so widely must have seemed, one must ask whether this transformation would have come about, at any rate in the way it did and in the time it did, without the allegations of public scandal.

The incident was salutary, and not just for the patients and staff at a special hospital containing a high proportion of patients with dangerous or violent propensities, in showing that a wider public can be aroused if there is any suspicion of brutality or lack of competence towards an especially ill-favoured group of

---

[49] *Report of the Review of Rampton Hospital* (Cmnd. 8073), HMSO, London, 1980, p. 19. The chairman of the Management Review Team was Sir John Boynton.

[50] L. Gostin, *Institutions Observed: Towards a New Concept of Secure Provision in Mental Health*, King Edward's Hospital Fund, London, 1986, p. 80.

people. From that response at least, we can take heart. Paradoxically, it is not the most dangerous, the most violent, or the most
severely sub-normal who are necessarily the most unrewarding
to treat. In a trenchant description, Dr Scott once spoke of the
'not nice' patients: 'the ones who habitually appear to be well
able to look after themselves but don't, and who reject attempts
to help them, break the institutional rules, get drunk, upset the
other patients, or even quietly go to the devil in their own way
quite heedless of nurse and doctor'.[51]

VI

To illuminate some of the issues that arise, we can now pause
here to consider the case of an individual offender whose
experience encapsulates many of the processes described in this
chapter. The story begins before prosecution and conviction and
continues after discharge from hospital. Usually far more
attention is paid to the sentence of the court, the instrument by
which convicted offenders are put into prison or hospital, than to
what happens when they come out. In the more serious cases,
when freedom is finally achieved it is often conditional—an intermediate status between liberty and custody, for a time at least,
when a released person's ability to survive in the community can
be tested and watched.

A quarter of a century ago, a man then in his mid-forties
became increasingly obsessed with the idea that his wife was
unfaithful to him and having affairs with other men. They had
been married for more than twenty years and had either ten or
eleven children; curiously, no one was quite sure how many. His
wife had also given birth to two stillborn children. The husband,
an ex-miner, had persuaded himself that some of the family had
been fathered by men other than himself and had rejected five of
his children, who had been adopted or informally cared for by
relatives and friends.

Brought up in a traditional mining village in the North of
England, he had left school at fourteen and went to work underground soon after. While still in his teens, he was signed off work

[51] Scott, *Has Psychiatry Failed?*, p. 8.

for a year on medical grounds, a contemporary report noting that there was a neurotic, phobic element to the breathlessness, exhaustion, and palpitations which he experienced. However, after an uneventful period of wartime service in the Army, he returned to the colliery and to work on the coal face. After some years he became afraid to go underground, fearing that something might happen, that the pit might cave in or the props might collapse. It was his nerves and stomach trouble, he said, that had finished him. He was now forty-two years of age. The Coal Board kept him employed on casual jobs above ground, but he was frequently absent sick, the medical reports referring to anxiety neurosis. From the time he gave up working below ground, his consumption of alcohol, already heavy at several pints of beer a day, increased.

The subject of this narrative, let us call him the self-deluded husband (SDH), had not behaved well towards his wife. Three times he had been before the Magistrates' court charged with assaulting her; three times he had been convicted and conditionally discharged or fined. On two other occasions he had incurred non-custodial penalties for causing wilful damage to property and being disorderly while drunk. Each of the offences had been preceded by sessions of heavy drinking. On his sixth court appearance, the third in less than two years, the Bench sentenced the SDH to one month's imprisonment after an assault causing actual bodily harm on a neighbour whom he accused of being the father of three of his children. Complaints to the police from the victim of this assault, and other men similarly accused, as well as, at intervals, from his wife and her relations, became commonplace, and unfortunately seem to have been treated as little more than a persistent nuisance—the by-products of just one more domestic dispute.

Only later, on his appearance in the Crown Court charged with double murder, did the full extent of the violence which had disfigured the SDH's married life come to light. It transpired that, shortly after the birth of their first child, nineteen years before, he had attempted to strangle his wife and had threatened with a bread knife his mother-in-law, at whose home they were living. Periods of living apart and reconciliations followed, but quarrels, threats, verbal aggression, and physical assaults characterized their married life. On his last but one appearance in the

Magistrates' court, the SDH was fined for having caused wilful damage to the front door of his mother-in-law's house, which he had tried to break down when his wife had left home and gone there, taking some of the children with her. A few months later he was committed to prison for non-payment of the fine, but after he had been in custody for about ten days his wife paid the fine and he was released. As they were returning together from the prison the SDH told her that while inside he had heard that the wife of a fellow inmate had been carrying on with other men. This appeared to prey on his mind, and from then on he frequently alleged that he was not the father of some or all of the children because at the time of their conception he was impotent. It was later established that he had in fact contracted gonorrhea during Army service, but apparently without any permanently harmful effects. The threats became more frightening, and on several occasions he threatened to kill his wife.

One evening, after being out drinking with a male neighbour whom he accused of being the father of the baby his wife was expecting, he returned home and assaulted his pregnant wife with the result that she received two black eyes. The next morning he went to the house of another neighbour, not his drinking companion of the night before, and told him that his wife had confessed to infidelity. She denied this, saying that she had never made such an admission, even under duress. She insisted that she had not had sexual relations with other men— indeed, that she had never been out with any man apart from her husband since their marriage when she was nineteen. And so it went on, in an escalating series of incidents—the wife being threatened and battered (on one occasion climbing out of a bedroom window and leaving the house during the night), and various men neighbours being accused—until the assault which led to the SDH going to prison for one month. On his release he was said to have told his wife that he had asked for his place to be kept warm because when he went back inside again it would be for something worth going back for.

After no more than a few weeks' co-habitation, his wife left home once more, taking only her youngest child, an infant, with her to her sister and brother-in-law who lived nearby. Five of the elder children, pathetic casualties in this tragedy, remained behind. The SDH found it hard to cope with looking after them, so he would take them round to his sister-in-law's house, only to have them sent back again by their mother. After this had happened a few times, he went around again, this time carrying a loaded shotgun. In the presence of his wife, he shot first her sister and then her brother-in-law at point-blank range. The brother-in-law died immediately from gunshot wounds to the head, while the sister died shortly after admission to hospital as a result of a severe haemorrhage from a fatal wound in her neck. His wife had run out of the house unharmed. After the killing, the SDH went to a drinking club. He had also been drinking elsewhere earlier in the day, although the part that alcohol played in the commission of the offence was never established. Still in possession of the shotgun and ammunition, and in a dishevelled and bloodstained state, he met his brother at the club. Later he was persuaded by his brother to go to their mother's home, where after a struggle, he was disarmed and the police informed. When arrested his first words were 'I've gone and bloody done it now, somebody will have to look after my kids ... you can stretch my neck tomorrow.'

Following examination on remand, medical opinion was unanimous that the self-deluded husband was suffering from a paranoid delusional psychosis. He continued to maintain that his wife had been persistently unfaithful and that the two victims had encouraged her in deceiving him. One of the doctors who examined him thought he would be a case for the insanity defence under the M'Naghten rules, but this view was not put to the court as the defence lawyers preferred to rely on the defence of diminished responsibility under Section 2 of the Homicide Act 1957. This was accepted by the court, and the defendant was found not guilty of capital murder on the grounds of his diminished responsibility, but was convicted of the manslaughter of both victims and sentenced to two concurrent terms of life imprisonment. The court may have considered the case too serious to warrant disposal by way of a hospital order, but in practice the treatment of mentally disordered offenders enables, although it does not ensure, movement from prison to hospital (and sometimes back again to prison) after sentencing. In this case the SDH spent twelve months under psychiatric observation in a prison hospital, where he continued to express paranoid delusional beliefs concerning his wife, although paradoxically he looked forward to and appeared to enjoy her visits. Within a year, his psychosis led to his transfer to a special hospital under the powers conferred on the Secretary of State by Section 72 of the Mental Health Act, 1959. On admission to Broadmoor he was calm, and able to give a clear account of his history and the offence, neither denying his guilt nor displaying any remorse for the killings. He repeated his conviction of his wife's infidelity towards him, and at a clinical conference the diagnosis of paranoid psychosis was confirmed.

Eight years of confinement and treatment passed. Apart from some stress at the falling off, and then the termination, of his wife's visits, with divorce and his wife's remarriage following some years later, the SDH's progress was uneventful. In the hospital setting his behaviour was reasonable and his manner cheerful, although nothing affected his apparently unshakeable delusion regarding his wife's infidelity and her family's connivance. He appeared twice before a Mental Health Review Tribunal, and on the second occasion was recommended for discharge to the home of a brother. Although his clinical condition was little altered, medical opinion inclined to accept his assertion that he no longer had any interest in his ex-wife and intended to make a new start in a different part of the country. A slight increase in emotional responsiveness and capacity for self-criticism was detected, and it was considered that, while it was doubtful if he would ever lose his fixation about his wife's infidelity, that alone was hardly sufficient to justify his continued detention. The release plan was later changed to allow him to go to a sister in the Midlands, whose home seemed to offer a more stable and supportive environment after a long period of detention. Supervision would be exercised by a senior social worker, and the Broadmoor authorities would keep a close eye on the case. Accordingly, having consulted the Lord Chief Justice, the Home Secretary authorized his conditional discharge from hospital. At the age of fifty-five, the self-deluded husband, whose delusion had long ago become an integral part of his personality, was once again at liberty, although the element of conditionality soon became relevant.

After a few months with his sister, the SDH formed a relationship with a sixty-three-year-old widow and moved in with her. This change in his situation required an alteration in the terms of his discharge warrant, which was approved. At first all seemed to go well, but then there were ominous reports that he was drinking heavily. Seventeen months after discharge, the taste of freedom ended: the woman with whom he was living went to the police. She complained of his heavy drinking and of having been beaten up and threatened with violence and disfiguration. The senior social worker had an alarming interview with the woman. The Home Office was informed that the SDH was temporarily in police custody on a charge of threatening behaviour, and that his recall was urgently recommended by the physician super-intendent at Broadmoor. Later the same day a warrant of recall was issued and handed to an escort from Broadmoor, who then collected the man from the police and returned him to the special hospital.

Reports from Broadmoor over the next long stretch described the SDH as morbidly paranoid, vulnerable to stress, and insight-less. After seven years he appeared once again before a Mental Health Review Tribunal, which did not feel able to recommend either discharge or a move to a local hospital at that stage. A year later, however, he was transferred to another special hospital, giving the same treatment and in conditions of similar security, but at least constituting a change of scene. When a further eighteen months had passed, it was unanimously agreed that he no longer needed the security of a special hospital, and ministerial agreement was obtained for transfer to a psychiatric hospital under the care of an experienced forensic consultant. By now he was gravitating back towards his roots, and the hospital was not far away from where several members of his large family were still living. They were contacted, and two of his daughters offered to help with his rehabilitation. Such magnanimity was not uniform, however. Not all families are united, especially one that had endured such agony. Another daughter wrote to her MP asking him to intervene, saying that she and her mother would be in fear of their lives if her father were to have any freedom in the locality. Further enquiries established that the family was deeply divided between those who supported, or at any rate tolerated, their father in the prospect of his discharge, and those who remained fiercely loyal and defensive of his ex-wife, who claimed never to have forgiven him for his crimes.

It was hardly surprising that, when the case was referred to the

Advisory Board on Restricted Patients at the Home Office, there were mixed reactions.[52] At the age of sixty-seven, it was thought that the SDH's propensities towards violence had probably declined sufficiently to make discharge for the second time an acceptable risk, although concern was expressed about the risks that might arise from excessive consumption of alcohol. The case was deferred for a detailed after-care plan to be prepared with specific provision for monitoring his drinking. A proposal for the SDH to go to a purpose-built residential hostel, not far from the special hospital, which was run by a trained psychiatric nurse and provided intensive care and rehabilitation, was subsequently approved. The Advisory Board underlined that alcohol would be the best indicator of potential dangerousness and recommended that any irresponsibility in this respect would argue strongly for recall to hospital. Home Office Ministers were consulted and six months' trial leave at the hostel was granted. Both the consultant psychiatrist and a senior social worker maintained close observation, and progress was sufficiently satisfactory for the leave period to be extended for further six-monthly periods.

One last hurdle remained. The SDH was still subject to a sentence of life imprisonment imposed all those long years ago in the Crown Court. Since his earlier discharge from Broadmoor, the case of any life sentence prisoner who has been transferred to hospital for treatment under the Mental Health Act is referred to the Parole Board before final release into the community on life licence.[53] A High Court judge will be among the members considering the case, as will a consultant psychiatrist. If the Board is unable to recommend release (in a case of this sort, generally on grounds of risk to the public), the Home Secretary has no power to release; if the recommendation is favourable, however, the Home Secretary is not bound to act upon the Board's advice. After nearly two years of living at the hostel, and twenty-two years since the day he took a shotgun to the house where his wife was sheltering and killed two of her relatives, the self-deluded husband's case came before the Parole Board. Now in his seventieth year, with his thoughts turning to

[52] Since 1985, cases of life sentence prisoners who have been transferred from prison to hospital in the course of their sentence are considered by the Parole Board, rather than by the Advisory Board on Restricted Patients.

[53] *Report of the Parole Board 1985* (HC, 428), HMSO, London, 1986, pp. 5–6.

the placid pleasures of pigeon racing, he had served twelve months in prison, seven and a half years at Broadmoor, seventeen months on conditional discharge, a further ten years after recall in two special hospitals, and almost two years under observation at a psychiatric hostel. The Board recommended his release on life licence with a condition of medical supervision by a consultant psychiatrist. An interval of eight months then passed while Ministers considered the case, consulting the judiciary again. Finally, the Home Secretary decided to accept the Parole Board's recommendation and authorized the release of the SDH under Section 61 of the Criminal Justice Act 1967.

Conditional freedom for a man who could never be free of the burden of his own past was attained once again, a fortnight after his seventieth birthday. Let us hope that this time it marks the end of a long road that has been uphill all the way.

VII

Some readers may be shocked by the implications of this example, showing how a mentally disordered offender can spend such a very long period in institutions of one sort or another. Others may feel that unacceptable risks are still being taken. Faced with the intractability of cases like this (and unfortunately, it is not unique), attempts to achieve well defined distinctions between penal sanctions imposed by the courts and the mental health powers of the state seem artificial. Incapacitation—that is, confinement to preclude the risk of further offences being committed in the community—tends to overshadow either sentencing considerations (punishment to fit the crime) or remedial treatment. The case history is also instructive in that it brings out the importance of the release procedures and the very detailed consideration they incur. The conditionality of discharge or release, the supervision that is exercised, and the liability to recall are crucial features which receive little public attention. Especially in serious cases of violence to the person, it is seldom a simple question of whether a man should stay in custody or be set at liberty; there are many steps between the two. Less serious crimes, including the great bulk of property offending, do not raise these questions in such an acute way, but even then, where the social evil lies more in the repeated nature of the offending than in the consequences to the individual victim, there are imponderable decisions to be made on the basis of risk. These are difficult issues extending well beyond mentally disordered offenders, and the implications for justice and public safety which are inherent in the concept of release on licence are taken further in Chapter 8.

Another case history, more briefly summarized, offsets the distressing story of the self-deluded husband. Once again, the offence was a killing within a family, although this time as a result of a depressive illness. Ever since the end of the Second World War, an ex-serviceman who had contracted syphilis had brooded on his condition and the medical treatment he had received. In 1945 penicillin was not available for the treatment of the disease. He was treated instead with arsenic and bismuth, but developed an allergy and did not receive penicillin for another five years. Before his marriage he had a thorough medical examination and had received a clean bill of health. He enjoyed a normal sexual relationship with his wife, and a son resulted from their union. As time went on the son married, the daughter-in-law became pregnant, and the birth of the first grandchild approached.

As the date of birth became imminent, the ex-soldier, referred to henceforth as the depressed grandfather (DG), became extremely depressed and was convinced that the child would be born a monster as a result of the disease he had become infected with so many years before. Obsessed by the desire that his wife should not witness this horror, and that he should not see her shamed, he took a hammer during the night and beat in his wife's skull, killing in her sleep the woman to whom he had been happily married for thirty years. Sensing that she was dead, he later said that he had made her more comfortable by crossing her arms over her chest and laying her on her back. He then swallowed a bottle of sleeping pills, washed down with whisky, and settled down on the bed beside her. After having nearly died from the suicide attempt, the DG was interviewed by the police in hospital and cautioned. Subsequently he was remanded for a short period in custody where he was examined by three doctors, one acting for the Crown. All were agreed that he was suffering from a mental illness too acute to allow him to remain in prison. He was therefore transferred to a regional secure unit as a condition of bail pending trial. There he was treated with electro-convulsive therapy and drugs. At the trial, the court found no difficulty in accepting a plea of not guilty of murder by reason of his diminished responsibility under Section 2 of the Homicide Act 1957, and ordered his detention at the regional secure unit under Section 37 of the Mental Health Act 1983 with a restriction order under Section 41 without limit of time.

After no more than a few weeks, however, it was possible to discontinue all medication as the DG was completely symptom-free. This was not unusual, since he was regarded as a classic case of depression and the duration of a depressive illness, properly treated, is often around eight weeks. One year later, having neither required nor received any further medication, the case of the DG came before a Mental Health Review Tribunal for the first time. Since the 1983 Act, tribunals have been chaired by lawyers with substantial judicial experience in the criminal courts, in this instance a circuit judge. The other two members were a consultant psychiatrist, who had examined the patient in advance of the hearing, and a lay member. The DG appeared in person and was legally represented by a solicitor. His son and daughter-in-law were present to support his application, and the responsible medical officer and a social worker from the regional secure unit were in attendance.

The question to be resolved by the tribunal was whether the DG's condition warranted his continued detention for medical treatment. The psychiatric evidence was unanimous that no further treatment was needed, although the Home Office took a more cautious view. In opposing the patient's suitability for discharge, the Home Office said that the serious and violent nature of the offence, and the comparatively short time spent in hospital, had led the Home Secretary to the conclusion that a longer period of sustained progress should be demonstrated before discharge could be safely considered.

A further complication was the unannounced arrival at the hospital where the tribunal was sitting of the patient's brother-in-law and his wife. It was the brother-in-law, living nearby, who had found the victim's body, the deceased being his sister, and he had neither forgotten nor forgiven. Faced with this unexpected development, the president of the tribunal thought it just to hear the brother-in-law, but in order to avoid confrontation did so in the absence of the patient and his son and daughter-in-law, who were also heard. This impromptu arrangement was agreed by the DG on the advice of his solicitor, who remained present to safeguard his client's interests. The brother-in-law spoke forcefully and with an unmistakable sense of grievance at the prospect of the DG's imminent discharge. He resented the fact that he had not been able to give evidence at the trial, and indeed had not been notified when the case had come before the court. In practice, the plea of guilty to manslaughter by virtue of diminished responsibility had precluded that opportunity, but the arguments advanced for better communication with witnesses in Chapter 2 are well illustrated by this case. The legal niceties had done nothing to shake off the brother-in-law's absolute conviction that his sister had been murdered by the DG. He thought it scandalous that discharge should be considered so soon and was convinced that medical opinion had been deceived. In his opinion, which was challenged by the testimony of the son and daughter-in-law, the DG would not be welcome to return to the local community and might be at risk of reprisals. As in the earlier case, it is entirely understandable how families can be so deeply divided in circumstances where such profound emotions are involved.

In reaching its decision, the tribunal courteously expressed the hope that the brother-in-law might come to accept that it had looked all the more carefully at the case because of the trouble he had taken to come and contribute to it. None the less, it saw no continuing ground on which the patient could lawfully be detained in hospital. The powers and obligations relating to the discharge of restricted patients in Section 72 of the Mental Health Act 1983 are precise. If a patient is no longer suffering from mental illness or any of the other forms of mental disability specified in the Act which make him liable to be detained in a hospital for treatment, then he is entitled to be discharged. The finding of the tribunal was that the medical evidence compelled the conclusion that the condition of the patient no longer warranted his continued detention for medical treatment, but that it was appropriate that he should remain liable to be recalled to hospital for such treatment should it be necessary in the future. The tribunal rejected the solicitor's application for absolute discharge, but decided that in law the applicant was entitled to conditional discharge. After consultation with the solicitor and social worker about arrangements for living accommodation, especially in view of the likelihood of the DG experiencing hostility from some quarters in the neighbourhood in which he had formerly lived, the direction was deferred for not more than two months. The conditions of discharge were that he should remain subject to social and psychiatric supervision and hospital recall. In the period of deferment the hospital social worker undertook to make arrangements for the element of social supervision after discharge to be provided by the probation service.

In this case, in marked contrast with the last, the procedures for the discharge of a mentally disordered offender who had responded satisfactorily to treatment resulted in a period of compulsory detention that was exceptionally short, compared with the consequences in terms of punishment and retribution, had the DG initially been channelled into the penal system as a result of receiving a life sentence, and then transferred to hospital for medical treatment. Although the case may be atypical, and few present such clear-cut examples of illness and cure, it is highly significant that considerations of tariff, that unyielding yardstick of retribution, were wholly absent.

VIII

Cloudy thinking, pragmatic judgments, and a pervasive sense of uneasiness characterize the approach towards the confused and bewildered battalions of mentally disordered persons who have committed crime. Most often they will be marched off to stand trial, but some will be detached from the main formation as lacking the competence to be tried. In offences of homicide, the more notorious the case, the greater the possibility of a conviction for murder and the mandatory life sentence. Medical evidence of abnormality of mind to an extent that substantially impairs a defendant's responsibility for his acts, if unanimous, will normally be accepted as sufficient grounds to reduce the charge from murder to manslaughter. But it does so only by recognizing that the mental capacity required to commit the crime of murder is greater than that required to commit the crime of manslaughter. A desirable enough aim, to give the court power to mitigate the sentence in compassionate cases, is achieved by devices that are hard to reconcile with rationality.

If the medical evidence for the defence is challenged by the prosecution, a series of unsatisfactory exchanges is likely to take place which demonstrates the incompatibilities between the legal and the medical approach. The inability of psychiatrists appearing as expert witnesses to communicate with lawyers, especially defence psychiatrists and prosecution counsel opposed to one another in an adversarial relationship, is evident for all to see. Dr Anthony Storr has argued that trying to define mental disorders and responsibility in ways that will satisfy a court is a fruitless task, since they are not susceptible to exact definition. Instead of their being called in the course of the trial as witnesses for the Crown or the defence, he would like to see psychiatrists employed as independent assessors to advise on the most appropriate disposal after guilt has been determined.[54] Although there may be procedural difficulties, this is an objective well worth aiming for. Storr recognizes there is a long journey ahead. But he is surely right to urge that the sooner it is embarked upon the better.[55]

Once the uncertainties of the trial are over, and guilt or innocence established beyond reasonable doubt (the words seem oddly ill-suited to the mental state of so many disordered offenders), further uncertainties loom. Is the case suitable for a hospital order, or does it call for a term of imprisonment? Would a non-custodial penalty, possibly involving supervision in the community, offer better protection for the public against the risk of future offending? On what grounds are the decisions to be made—remedial or penal? Should the practical consequences of the shortage of beds in hospitals for patients requiring to be held in conditions of security play so large a part? (It is hard to see how it can be ignored.) Should medical (but essentially non-diagnostic) distinctions, made primarily for the purposes of legal categorization, lead to such striking discrepancies in their outcome? These are questions for the doctor as well as the judge to answer. It is important that there should still be an

[54] Masters, *Killing for Company*, includes a postscript by Anthony Storr, FRCP, FRCPsych., pp. 317–25.

[55] Ibid., p. 325.

opportunity to examine psychiatric witnesses, but it would do much to improve communication and lead to the emergence of a more rounded view if this were to be done after the facts have been established, but before sentence is passed. This already happens when the defendant pleads guilty, but not in contested cases, where counsel seek to discredit expert witnesses in order to impress the jury. In homicide cases, many juries faced with deciding the crucial question of whether or not the defendant's mental state was substantially impaired must find the exchanges between lawyers and psychiatrists contradictory and confusing. It would be a constructive step towards a less adversarial relationship if the deployment of expert witnesses in contested trials of mentally disordered offenders were to be subjected to a thorough independent review. The present practice is not merely humiliating for the doctors involved, and unhelpful to the jury, but falls some way short of the standard of justice, imperfect as it is, that could be attained in these most difficult cases.

The link connecting mental instability and criminal guilt will never be an easy one. The human mind is too profound and too mysterious for there to be any prospect of absolute reliability in diagnosis and explanation of past actions or prediction of future conduct. This is especially true when the main source of information is what the offender has told the psychiatrist, rather than a result of empirical observation. In penology a place must be reserved for non-professional wisdom, common sense and knowledge of human nature, alongside the specialist skills of the psychiatrist and psychologist. It may seem tame as a conclusion, but I have few nostrums to offer beyond those implicit in this sequence. Study closely the causes of all forms of deviance where a crime has occurred. Follow it up; try to find out who did what to whom. Resolve disputed questions of fact in a court of law. After guilt has been established, assess the offender's mental state from the standpoint of whether he requires and is likely to respond to remedial treatment. If he does then he should follow the hospital route; if not, the road must lead to prison. Be humane, not sentimental. Never give up.

5

The Penalty for Murder

I

The penalty for murder is prescribed by Parliament. At the end of the trial, if the defendant is found guilty, the judge pronounces the sentence required by law. He does not pass sentence in the usual way, nor has he any discretion to vary the mandatory sentence of life imprisonment. This procedural distinction underlines the symbolic quality of the penalty for murder. The deliberate taking of another life has been regarded as so heinous a crime that it should be marked by a unique penalty to demonstrate society's condemnation of such grave criminal conduct. For just under a century—from 1861, when for all practical purposes murder became the only offence to continue to attract the death penalty,[1] until the Homicide Act 1957—the mandatory penalty for murder was death by hanging. Capital punishment lingered on for certain categories of murder for a few years more, until abolition in 1965.

Note that the uniqueness attaches to the mandatory character of the penalty rather than to the life sentence or the dreadful consequences of the crime itself. Murder is not the only crime that results in the premature termination of another human life. Manslaughter and infanticide (the killing by a mother of her newly born child) are distinct offences, although with similarly fatal results for the victim. Neither attracts a mandatory penalty; in fact, both owe their status as lesser offences to the necessity of finding ways to mitigate the unyielding severity of the sanction for murder. The motor car too is an instrument of carnage. The most serious of all motoring offences—causing death by reckless driving—was first separated from other forms of manslaughter in 1956. Until 1977 the offence was 'causing death by dangerous

---

[1] L. Radzinowicz, *A History of English Criminal Law and its Administration from 1750,* vol. 4, Stevens, London, 1968, p. v.

TABLE 6. *Homicides, by offence, 1981–5*

Offence	No. of persons convicted in				
	1981	1982	1983	1984	1985
Murder	126	184	132	156	173
Section 2 manslaughter (diminished responsibility)	84	90	84	72	65
Other manslaughters	167	182	164	168	200
Infanticides	11	3	9	7	3
Causing death by reckless driving	187	220	190	163	216

driving' and, as the figures in Table 6 show, this form of homicide has led to more convictions than any other category resulting in death. It also needs to be remembered that these convictions account for only a small proportion of the 5,000 or so lives lost each year in motor vehicle traffic accidents.[2] The punishment of motorists is referred to later in this chapter.

When a person is on trial for murder, the emphasis is often not so much on the deed itself (what happened and who did it), as on questions of intent (what was in the defendant's mind at the material time), and the extent of his responsibility for his action. Because so many murders stem from domestic quarrels or breakdowns in personal relationships, a large number of defendants plead either guilty, to the substantive murder charge, or not guilty to murder by virtue of diminished responsibility but guilty of manslaughter. The burden of guilt and remorse in many of these cases can be so self-destructive that each year a substantial proportion of all homicide suspects commit suicide before facing trial.[3] It is easy to get swamped by homicide statistics, but one stands out above all others. This is the fact that, in about three-quarters of all offences recorded by the police as homicide, the

[2] Deaths caused by motor vehicle traffic accidents were 5,059 in 1983, 5,102 in 1984, and 2,511 for the first six months of 1985. *OPCS Monitor*, 15 Oct. 1985.

[3] The Royal Commission on Capital Punishment found that 29.1% of murder suspects had committed suicide between 1900 and 1949, and for 1950–9 the proportion had risen to 33%. (See L. Blom-Cooper and T. P. Morris, *A Calendar of Murder*, Michael Joseph, London, 1974, p. 278.) More recent homicide statistics suggest a lower proportion, possibly as a consequence of the increased incidence of murder.

victim was already acquainted with the principal suspect. In nearly half, the victim was a member of the suspect's family, or was a co-habitant or lover, while the number of murder victims who were sons or daughters was higher in 1985 than in any year between 1975 and 1984.[4]

The essentially intimate, even domestic, nature of most murders is not widely appreciated. Murder is a crime committed by women as well as men, and by a sizeable group of young people below the age of eighteen. Often the offenders will be people of previously good character whose self-control has given way in circumstances that reveal all the unpredictability and frailty of human nature. There is, for example, the immature and inexperienced youth who kills because he has been taunted about his sexual performance; there are friends who get into fights, and lovers of both sexes who kill their partners; there are marriages from which all love and affection has long since departed which end in this culminating tragedy; and of course, endemic to our times, there are the murders that are not planned in advance but result from the misuse of alcohol or drugs. In the south of England there is a complete prison, and a very well run one, which is occupied entirely by life sentence prisoners, the great majority of whom have killed within an existing relationship with the victim and have had no previous experience of prison. Although governed by a woman, this prison contains only male inmates. However, there is food for thought in the statistic that at the end of 1986 there were seventy women serving life sentences. This compares with a total life sentence population of 2,250 at 31 October 1986.[5]

The numbers of men, women, and young persons convicted of murder over the last five years and sentenced to life imprisonment has been fairly stable, as the statistics in Table 7 show. Young persons under the age of eighteen at the time of the murder are sentenced to be detained at Her Majesty's Pleasure for an indefinite period under Section 53 of the Children and Young Persons Act 1933.

[4] These statistics relate to offences that were initially recorded by the police as homicide in 1985 and were still classified as such in September 1986. See *Criminal Statistics England and Wales 1985* (Cm. 10), HMSO, London, 1986, p. 56 and Table 4.4 on p. 61.

[5] *Parl. Debates*, HC, 107 (6th ser.), col. 654, 18 Dec. 1986. The total lifer population is estimated to increase at about 80–100 annually.

TABLE 7. *Convictions for murder, by age and sex, 1981–5*

	1981	1982	1983	1984	1985
Murder convictions					
Mandatory life sentences					
Male	107	147	112	126	148
Female	4	7	4	8	7
TOTAL	111	154	116	134	155
Young persons under 18 convicted of murder					
Male	15	29	15	22	16
Female	—	1	1	—	2
TOTAL	15	30	16	22	18

Lacking in homogeneity as an offence, murder covers a wide variety of circumstances and vastly differing degrees of culpability. While a clear majority of murders can be regarded as falling into the domestic category, there are others which are of exceptional gravity and to a greater or lesser extent represent an overt threat to society. Armed robbers may be willing to use a gun to kill if it proves to be necessary in securing a financial objective; police or prison officers may lose their lives in the course of performing their duties; terrorists may bomb or shoot indiscriminately in the pursuit of political objectives; and vulnerable children may be sexually abused or battered to the point of death. For murderers of this sort, there is little public or judicial sympathy. In pronouncing the sentence of life imprisonment, the judge may add a recommendation as to the minimum period of years that should elapse before a prisoner is released on licence, and it is rare for the Home Secretary not to follow such a recommendation.

The release of any life sentence prisoner requires the authority of the Home Secretary. His discretion is subject to a favourable recommendation by the Parole Board after it has considered the prisoner's case, and he is also required by statute to consult the Lord Chief Justice and the trial judge, if still available, before authorizing release. In November 1983 the then Home Secretary, Leon Brittan, announced that, taking account of public concern about violent crime, he intended in future to exercise his discretion in such a way:

that murderers of police or prison officers, terrorist murderers, sexual or sadistic murderers of children and murderers by firearm in the course of a robbery can normally expect to serve at least 20 years in custody; and there will be cases where the gravity of the offence requires a still longer period. Other murderers, outside these categories, may merit no less punishment to mark the seriousness of the offence.[6]

II

Whereas reports of murder trials usually give prominence to the more lurid details of the killing, the real issue before the court when a defendant pleads not guilty may relate to his intentions.[7] For murder to be established, it is not necessary for a deliberate intent to kill to be proved beyond reasonable doubt. In the more straightforward cases this may be so, but in others evidence of an intent to cause serious bodily harm will be enough to convict of murder. But then the sky darkens, and other factors—notably, foresight, or the lack of it—cloud the outlook still further. In a leading case in 1974, the House of Lords expressed some diverse views on the degree of foresight required. The circumstances were poignant; the appellant, a woman, had started a fire in a house for the purpose of frightening another woman in the house into leaving the neighbourhood. In the course of the fire the children of the other woman lost their lives.[8] Five years later the Criminal Law Revision Committee summarized the effect of this decision on the law of murder. The Committee concluded that it would be murder if a person killed by doing an act:

(i) intending to kill; or
(ii) intending to cause serious bodily harm; or
(iii) knowing that death is a [highly] probable result of the act; or
(iv) knowing that serious bodily harm is a [highly] probable result of the act [provided that the act is 'aimed at' someone][9]

[6] *Report of the Parole Board 1984* (HC 411), HMSO, London, 1985, p. 23. The lawfulness of this policy was challenged in the High Court by way of judicial review, but was upheld in the House of Lords (ibid., pp. 5–6).
[7] Intention and *mens rea* in murder are discussed by Dr Anthony Kenny in *The Ivory Tower,* Basil Blackwell, Oxford, 1985, pp. 16–30.
[8] *Hyam* v. *DPP* [1975], AC 55.
[9] *Criminal Law Revision Committee, 14th Report: Offences against the Person* (Cmnd. 7844), HMSO, London, 1980, p. 8.

The words in brackets were used by the Committee to indicate lack of unanimity or dissent among the law lords. Later interpretations by the appellate courts in 1985 caused the *Criminal Law Review* to comment that the respective ambits of murder and manslaughter had expanded, contracted, and expanded again in the manner of a piano accordion, with the higher judiciary in the House of Lords, the Privy Council, and the Criminal Division of the Court of Appeal playing the tunes.[10] As a result of these cases, the present state of the law is that the word 'highly' should be deleted and that, for the time being at least, juries will have to decide whether or not a defendant charged with murder foresaw death or really serious harm as being the probable or the natural consequence of his act.[11] The point is of some importance to those accused of the gravest offence known to the criminal law, since some judges believe that the addition of the word 'highly' would be too favourable to the defendant and that an awareness of ordinary probability or likelihood should be sufficient.

An illustration of the fact that these issues were no legal quibbles but went to the heart of criminal offending and personal responsibility occurred in the course of the year-long miners' dispute in 1984/5. Of the many incidents of violence that took place,[12] the most serious in its consequences resulted in two Welsh miners being found guilty of the murder of a taxi-driver taking a miner to work in defiance of his union's instructions. The defendants strongly objected to the working miners' conduct, but claimed that their action in pushing two blocks of concrete from a bridge on to the road as a convoy of vehicles was passing below was simply intended to frighten the occupants of

[10] 'Murder and Manslaughter: The Case for Statutory Reform', *Criminal Law Review* (1986), 1–2.

[11] The cases referred to in the *Criminal Law Review's* editorial (ibid.) were those of *Leung Kam-Kwok* v. *R.* (1984) 81 Cr. App. R. 83, P.C.; *R.* v. *Moloney* [1985] AC 905; and the Court of Appeal's judgment in *R.* v. *Hancock and Shankland* [1985] 3 WLR 1014. The latter case went to the House of Lords on appeal by the Crown on a point of law.

[12] According to one estimate, in the course of the dispute more than 10,000 criminal charges were brought and some 1,400 police officers injured. See J. Croft, *Confidence in the Law*, Conservative Study Group on Crime, published Report no. 1, London, 1986, p. 6.

the taxi or to cause an obstruction.[13] They denied any intention to kill the driver or his passenger, or to cause them serious injury. Nevertheless, by a majority of 10 to 2 a jury found that the accused should have foreseen that the natural consequence of their act was likely to cause death or serious injury. Accordingly, they were convicted and sentenced to life imprisonment, as the law required.

The political reaction in South Wales, as might be expected at a moment when feelings were running high, was vocal and hostile. Local MPs condemned the dangerous and irresponsible incident that led to the death, but dissented strongly from the view that it was a proper case of murder, arguing that no such charge should ever have been brought. Although their grasp of the law on murder may have been shaky, their view of events prevailed in the Court of Appeal. In his judgment the Lord Chief Justice, Lord Lane, detected some ambiguity in the meaning of the words 'natural consequence' in the trial judge's direction to the jury and allowed the appeal, although not without some hesitation. In an attempt to clarify matters for the future, the court laid down guidelines on questions to be put to the jury where the defendant's motive was not primarily to kill or injure. Verdicts of manslaughter were substituted for the murder convictions, and sentences of eight years' imprisonment were imposed on each of the appellants.

In the House of Lords, Lord Scarman set the case in the context of public policy. Crimes of violence, he said, where the purpose is by open violence to protest, demonstrate, obstruct, or frighten were on the increase:

Violence is used by some as a means of public communication. Inevitably there will be casualties: and inevitably death will on occasions result. If death results, is the perpetrator of the violent act guilty of murder? It will depend on his intent. How is the specific intent to kill or to inflict serious harm proved? Did he foresee the result of his action?

[13] In the light of the widespread public condemnation of the irresponsibility of the defendants' conduct, irrespective of whether it amounted to murder or manslaughter, it seems hardly credible that the same actions could recur in the furtherance of an industrial dispute. Yet in 1986 police reported that missiles had been thrown from a bridge over the A2 at Dartford as lorries on their way to collect newspapers from the Wapping plant of News International were passing below, and that one driver had suffered an eye injury. *Daily Telegraph,* 9 Dec. 1986.

Did he foresee it as probable? Did he foresee it as highly probable? If he did, is he guilty of murder?[14]

The House of Lords did not dissent from the conclusion reached in the Court of Appeal, but it disapproved of the use of guidelines, including some formulated in an earlier case in the House of Lords on which the trial judge had relied, as well as those promulgated by the Lord Chief Justice on appeal. Juries in murder trials, Lord Scarman said, should be encouraged to use their common sense in reaching a decision on the facts. It was the duty of the judge to direct the jury in law and to help them focus upon the particular facts of the case. Guidelines were not rules of law and should be used only sparingly, and be limited to cases of real difficulty. In general, the House of Lords regarded guidelines in this field as being neither wise nor desirable.

Quite apart from the political overtones, the case of the Welsh miners took its place as one more in a long line of instances where confusion as to what constitutes the law of murder and the necessary degree of foresight of consequences has led to doubt and differing interpretations. The search for greater clarity and precision in the definition of the gravest crime in the criminal calendar is not helped by the fact that murder is still a common law offence, nowhere defined on the statute book, although several proposals for its enactment have been made. Little if anything has changed since the Criminal Law Revision Committee in 1980 commented that the present law was in need of clarification to remove doubts and was 'capable of substantial improvement'.[15]

Manslaughter cases, too, are riddled with tricky questions of intent, especially in the area of recklessness and negligence. It is for reasons of this sort that some lawyers would like to see a thorough overhaul of the law on offences against the person. The Criminal Law Revision Committee put forward a comprehensive scheme for reform in its 1980 report, and in 1985 the Law Commission published as a discussion paper a scheme for codifying the criminal law which adopted the Criminal Law Revision Committee's proposals relating to offences against the

---

[14] *R. v. Hancock and Shankland* [1986] 1 AC 455, at p. 468.
[15] *Criminal Law Revision Committee 14th Report* (Cmnd. 7844), p. 9.

person.[16] Although these proposals maintained the distinction between murder and manslaughter as separate crimes, something which is considered later in this chapter, it would nevertheless be a step in the right direction towards making this crucially important branch of the law comprehensible and accessible to those who are likely to be affected by it and have so much to lose. These are worthwhile benefits to aim for. The alternative is to scrap the distinction between murder and manslaughter altogether. Statutory reform in this field would not only bring greater certainty to a perplexing limb of the criminal law, but at the same time would provide a better defined context for the defences of diminished responsibility and provocation which relate solely to homicide cases. Is it only laymen who see a paradox in defining by statute defences to a crime that is itself lacking statutory definition? Few Home Secretaries have relished the prospect of legislating to reform the law on offences of violence against the person, largely because rational consideration of the law on homicide has been haunted by the emotional, and enduringly political, spectre of the death penalty for murder.

III

In the two decades that have passed since the abolition of capital punishment, there have been regular tests of opinion in the House of Commons. Each new Parliament has seen a debate on a free vote, in which points of view are strongly expressed and voting crosses party lines. Each time there has been a majority against restoring capital punishment, and each time the abolitionists fervently hope that the issue, one of the most contentious in contemporary politics, will finally be laid to rest. No doubt Whips and party managers, their hearts set on accomplishing or thwarting the Government's legislative programme, echo the sentiment. Dormant though the possibility of legislative change may be for the duration of one more Parliament, the intensity of interest in capital punishment shows little sign of abating, and the merest spark can set passions ablaze once more.

The fact that some of the debates on capital punishment,

---

[16] *Codification of the Criminal Law: A Report to the Law Commission*, Law Com. no. 143 (HC, 270), HMSO, London, 1985.

one in July 1983 being an example, have followed hard on the heels of a general election means that members return to Westminster, some having been elected for the first time, with the vehemently expressed opinions of the hustings still ringing in their ears. Moreover, enquiries may have been made and understandings sought before selection and adoption in order to ensure that the local party does not find itself committed to a standard-bearer who can be depicted as being soft on crime. This makes it harder for an MP to be his own man on arrival in the House of Commons. Edward Heath acknowledged this in the 1983 debate, recognizing that there are many pressures on how MPs should vote. But at the end of the day, he declared, each Member of Parliament must use his own judgment: 'It is the basis of the British constitution; we are not mandated, we cannot be mandated by a selection committee, by a constituency committee or by the public as a whole.'[17]

Parliamentary debates on the death penalty are prone to be dominated by what is most in the news at the time. In 1983 it was terrorist killings that tended to overshadow everything else. Yet the enduring question for MPs is that of public safety. Is capital punishment capable of providing society with a protection that is not afforded by other forms of punishment? If this can be established, then is it not the duty of Parliament to provide the maximum degree of protection for the individual citizen? This is a clear and reasonable test, although one that produces conflicting answers. Those favouring the restoration of the death penalty point to the fact that the number of homicides (i.e. all killings, and not simply murders) has nearly doubled since the abolition of capital punishment. Armed robberies involving the use of firearms have increased even more dramatically, from 464 in 1969 to 1,893 in 1981. It is at the core of the restorationists' case that, if there is no death penalty to frighten the criminal, he may go out with a gun and kill as a result. On the other hand, if the sanction of capital punishment existed, the same person might calculate the chances of being arrested, charged, convicted, and ultimately brought to the point of sentence knowing that he would have to face the death penalty; in such circumstances he would be likely to conclude that the risks were

[17] *Parl. Debates,* HC, 45 (6th ser.), col. 902, 13 July 1983.

too high and that he would not consequently be prepared to use a gun in the commission of a robbery. The fear of death is more powerful than any other deterrent, it was claimed by Sir Edward Gardner QC, the mover of the motion and a veteran campaigner for capital punishment, and is a fear that can make ordinary, normal human beings behave differently.[18] Arguments on these lines depend more on belief and on instinct than on statistics, but they are strongly felt and evoke widespread popular support inside and outside Parliament.

Abolitionists reject the assertion that, by reintroducing the death penalty, the murder rate and the number of violent crimes, including those where firearms are used, would be reduced. They reason that the upward trend in the number of homicides started in 1960, five years before abolition, and followed a fifteen-year cycle when the trend was generally downward. In what was intended to be a dispassionate survey of official statistics, the newly appointed Home Secretary, Leon Brittan, told the House of Commons that there were forceful arguments against accepting the rise in the number of homicides since abolition as retrospective proof of capital punishment's deterrent effect. He pointed out that the rise in crime, particularly violent crime and not just homicide, had been a general phenomenon since the early 1960s. If the ten-year period 1963–72 was compared with the subsequent ten years, it was evident that the number of homicides rose by 45%, whereas the number of other serious offences of violence against the person rose by 49%.[19] While no evidence of a uniquely deterrent effect of capital punishment can be substantiated on these figures, the extent to which abolition may have had an impact on the willingness of those involved in robberies or other violent crimes to carry guns, and to use them if necessary to avoid arrest, is unquantifiable.

Behind the unresolved utilitarian controversy on deterrence lie deeply rooted moral attitudes. These have been deployed more persistently and persuasively over the years by the abolitionists, although this may be changing as the influence of the new Right expands from economic policy into social policy. Emphatic voices can now be heard insisting that some murders are so hideous that the execution of the offender is the only morally

---

[18] *Parl. Debates,* HC, 45 (6th ser.), col. 883.   [19] Ibid., col. 887.

appropriate response. Justice, they say, requires that every crime deserves retribution proportionate to its gravity and to the culpability of the criminal. The death penalty is warranted as no more than the gravest form of socialized vengeance for the gravest offences. To punish accordingly is to do justice.[20] In this way retribution, or what is currently described in America as 'just deserts for past crimes', is distinguished from revenge in terms of the satisfaction of vindictive feelings. Only the dimmest outlines of this philosophy were evident in the House of Commons debate in July 1983, although all references to retribution evoked a vociferous response.

In contrast, the liberal morality was articulated on both sides of the House. Notable Tory abolitionists, including Edward Heath and Norman St John-Stevas, were aligned with Opposition front benchers Roy Hattersley and Merlyn Rees, and with Roy Jenkins, twice Home Secretary, for the Social Democrats. The first speaker from the Labour side, Roy Hattersley, opened with an eloquent credo:

I am wholly and irrevocably opposed to the re-introduction of capital punishment. I am opposed in principle because I believe that to legislate for the judicial execution of a man or woman held in the state's safe custody would be a reversion to barbarism. We would, in this country, become the only Western democracy where the state possessed and exercised the right to kill as judicial punishment. Nothing can justify savagery of that sort. A reversion to such a practice would debase and, in the literal sense of the word, demoralize us all.[21]

The motion for debate, a straightforward call for the restoration of the death penalty for murder, was obscured by a cluster of amendments seeking the reintroduction of capital punishment for specific forms of murder. One of these—the imposition of the death penalty for terrorist murders—surprisingly gained the support of the Home Secretary. In his speech, Leon Brittan came down in favour of the amendment, not on grounds of deterrence, since he accepted that the true fanatic is not deterred by the possibility of execution, some indeed being attracted by the prospect of martyrdom, but as an unambiguous signal of the total and absolute repugnance of the state for those whose crimes are

[20] E. van den Haag, *Deterring Potential Criminals* (Social Affairs Unit, Research Report 7), London, 1985, p. 4.
[21] *Parl. Debates,* HC, 45 (6th ser.), cols. 892–3, 13 July 1983.

aimed at undermining its foundations. Subsequent speakers laid stress on the difficulties of defining terrorism (even membership of the IRA and other proscribed organizations cannot always be proved in the courts north or south of the border), and in particular the unacceptability of the death penalty being imposed by judges sitting without juries in Northern Ireland.

The opponents of the policy of executing terrorists had the stronger case. In the heightened atmosphere that would accompany a capital charge, it is virtually unthinkable that jury trials could be held in Northern Ireland. Intimidation and partiality would be unavoidable. If convictions were obtained in non-jury courts, perhaps composed of three judges sitting together, and executions followed, tension and disorder would mount, erupting into communal violence. Once again the tribal blood sacrifice would be celebrated and new martyrs made, taking their place in a sequence of mythical patriots immortalized as having given their lives to the cause. Nor would the Republicans have the field to themselves; para-military Loyalists would soon be lining up for a place in their Protestant pantheon. If the courts were unable to convict on grounds of lack of evidence or insufficient proof, acknowledged terrorists would be acquitted, returning as heroes to their divided communities. Either way, the political consequences of this manner of expressing the repugnance of the state for acts of terrorism would be disastrous.

It should be noted that not all of the politically motivated incidents of bombing and shootings that have taken place on the British mainland in recent years have been related to the conflict in Northern Ireland. Arab and other terrorist movements have extended their disputes to Britain, with the result that between 1977 and 1983 eleven people lost their lives at the hands of non-Irish terrorists. Yet the inability of capital punishment to deter those who are politically motivated and ready to risk their lives whatever the cause, as well as unforeseeable political repercussions and reprisals that would ensue, make the death penalty equally self-defeating in these cases.

When the votes were counted at the end of the debate after a winding-up speech by a formidable Tory polemicist, Teddy Taylor, the amendment on terrorism was defeated by 361 votes to 245, the main motion being lost by a slightly larger majority of

368 to 223. Despite the proximity of the general election and a large influx of about one hundred new Conservative members, the majority against restoring capital punishment was remarkably close to what it had been on previous votes. Edward Heath's exhortations had apparently been heeded. Shortly after the 1979 general election, in July of that year, a similar motion to reintroduce capital punishment had been rejected by 362 votes to 243.

Table 8, overleaf, shows the results of all votes in the House of Commons on capital punishment between 1955 and 1987. In the Lords the issue has been less frequently raised; the last full-scale debate on the death penalty (not solely for offences of terrorism) was held in December 1969.[22] When the Homicide Act 1957, which had retained capital punishment for certain categories of murder, was repealed in 1965, Parliament decreed that the period of abolition should lapse on 31 July 1970 unless affirmative resolutions of both Houses were passed in favour of extending abolition, either indefinitely or for a further trial period. Both Houses voted in December 1969 to continue abolition indefinitely, in the Commons by a majority of 343 to 185 votes, and in the Lords by 220 to 174 votes.

IV

In spite of these consistent parliamentary majorities, and the reality that capital punishment has been obsolete for over twenty years, the controversy continues. Strong feelings lie latent and from time to time are activated. Events are the spur. Popular opinion is easily aroused, particularly by press reports of murders of exceptional violence or depravity. The public reaction is not confined to sympathy towards the victims. Sensationalism apart, a genuine sense of outrage and hostility may spread so that normally tolerant and unexcitable individuals are moved to demand the most extreme penalties for offenders, even before the full facts are known. Advance publicity can only make it harder for accused persons to get a fair trial, although juries have consistently demonstrated they can make up their own minds when the law has been explained to them. Dramatic

[22] *Parl. Debates,* HL, 306 (5th ser.), cols. 1106–1322, 17–18 Dec. 1969.

TABLE 8. *Votes in the House of Commons on capital punishment since 1955*

1. Motion on Royal Commission on Capital Punishment, 10 February 1955: *approved* 245–214.[a]

2. Death Penalty (Abolition) Bill 1955/6. Second Reading, 12 March 1956: *approved* 286–262.[b]

3. Death Penalty (Abolition) Bill 1955/6. Third Reading, 28 June 1956: *approved* 152–133.[c]

4. Homicide Bill 1956/7. Third Reading, 6 February 1957: *approved* 217–131.[d]

5. Murder (Abolition of Death Penalty) Act 1965. Vote on Second Reading: *approved* 355–170; vote on Third Reading: 200–98 *against* capital punishment, 13 July 1965.[e]

6. Murder (Abolition of Death Penalty) Act 1965, (Amendment) Bill Vote, 24 June 1969. Motion for leave to bring in Bill: *defeated* 126–256.[f]

7. Debate on Motion that 1965 Act should not expire, 16 December 1969. Motion to continue abolition: *agreed* 343–185.[g]

8. Criminal Justice Bill 1971/2. Standing Committee G. New clause to bring back capital punishment for certain forms of homicide: *defeated* 7–10, 9 March 1972.[h]

9. Penalties for Murder Bill 1972/3. Leave to introduce: *refused* 78–50, 26 January 1973.[i]

10. Restoration of Capital Punishment Bill 1972/3. Leave to introduce: *refused* 320–178, 11 April 1973.[j]

11. Northern Ireland (Emergency Provisions) Bill 1972/3. Abolition of capital punishment in Northern Ireland: *approved* 253–94, 14 May 1973.[k]

12. Motion disapproving the reintroduction of capital punishment for crimes of terrorism: *approved* 369–217, 11 December 1974.[l]

13. Motion on Terrorist Offences (Penalty), 11 December 1975. Motion in favour of capital punishment: *defeated* 232–361.[m]

14. Motion to reintroduce capital punishment, 19 July 1979: *defeated* 243–362.[n]

15. Criminal Justice Bill 1981/2. Report Stage, 11 May 1982. New clauses moved seeking to reintroduce capital punishment: *defeated* 195–357 (c. 682) (murder); 208–332 (c. 686) (terrorism); 176–343 (c. 691) (firearms or explosives); 208–332 (c. 694) (murder of police or prison officers); 151–331 (c. 698) (murder in course of robbery or burglary with offensive weapons).[o]

16. Motion to restore the death penalty for murder, 13 July 1983: *defeated* 223–368. Amendments also *defeated*:

   (a) murder resulting from acts of terrorism: 245–361 (c. 962);
   (b) murder of a police officer during the course of his duties: 263–344 (c. 967);
   (c) murder of a prison officer during the course of his duties: 252–348 (c. 971);
   (d) murder by shooting or causing an explosion: 204–374 (c. 975);
   (e) murder in the course or furtherance of theft: 194–369 (c. 978).[p]

17. Armed Forces Bill 1985/6. New clause to abolish capital punishment for certain military offences: *defeated* 38–113, 10 April 1986.[q]

18. Criminal Justice Bill 1986/7. Third Reading, 1 April 1987. New clause on the death penalty: *defeated* 230–342.[r]

[a] *Parl. Debates*, HC, 536 (5th ser.), col. 2180, 10 Feb. 1955.
[b] *Parl. Debates*, HC, 550 (5th ser.), col. 146, 12 Mar. 1956.
[c] *Parl. Debates*, HC, 555 (5th ser.), col. 838, 28 June 1956.
[d] *Parl. Debates*, HC, 564 (5th ser.), col. 566, 6 Feb. 1957.
[e] *Parl. Debates*, HC, 716 (5th ser.), col. 464, 13 July 1965.
[f] *Parl. Debates*, HC, 785 (5th ser.), col. 1232, 24 June 1969.
[g] *Parl. Debates*, HC, 793 (5th ser.), cols. 1148–1298, 16 Dec. 1969.
[h] *Parl. Debates* (Commons), SC (1971–2), G, cols. 860–967, 7 and 9 Mar. 1972.

[i] *Parl. Debates*, HC, 849 (5th ser.), col. 896, 26 Jan. 1973.
[j] *Parl. Debates*, HC, 854 (5th ser.), col. 1340, 11 Apr. 1973.
[k] *Parl. Debates*, HC, 856 (5th ser.), col. 1142, 14 May 1973.
[l] *Parl. Debates*, HC, 883 (5th ser.), col. 636, 11 Dec. 1974.
[m] *Parl. Debates*, HC, 902 (5th ser.), col. 724, 11 Dec. 1975.
[n] *Parl. Debates*, HC, 970 (5th ser.), col. 2122, 19 July 1979.
[o] *Parl. Debates*, HC, 23 (6th ser.), cols. 608–701, 11 May 1982.
[p] *Parl. Debates*, HC, 45 (6th ser.), cols. 882–986, 13 July 1983.
[q] *Parl. Debates*, HC, 95 (6th ser.), col. 441, 10 Apr. 1986.
[r] *Parl. Debates*, HC, 113 (6th ser.), col. 1191, 1 Apr. 1987.

Source: House of Commons Library (Research Division).

trials are followed with a gruesome interest that is distasteful or worse. Is it possible to see in the near-obsessive attention paid to reported crimes of murder, the detection of the perpetrator, and the subsequent trials a reflection of the cathartic functions served by the earlier and more ghastly spectacles of public executions?

The place accorded to murder in the consciousness of the British public is hard to assess. Feelings of revulsion are similarly evoked by particularly serious injuries inflicted in disturbing incidents of child abuse, rape, and assaults on elderly people which do not result in the victim's death. The popular response is one of condemnation, an absolute rejection of the behaviour of the perpetrator linked with a demand for his punishment. None of this is peculiar to murder. The well-worn vocabulary of Fleet Street—'brutal', 'vicious', 'appalling', 'sickening' and 'shocking'—is in constant use to describe crimes of violence of all sorts. Does such a consistently high level of media exposure blunt humanitarian susceptibilities, or does it serve to keep a collective social conscience attuned to the sufferings of fellow citizens who have been the victims of crime? Does the uniformity of response, with its unvarying tone of outrage and denunciation, reinforce prejudice, or is it an authentic expression of the mores of a closely knit and like-minded society? These are questions to which there are no simple answers—indeed, probably no answers at all—but they indicate the parameters within which policies towards crime are contained.

In an indefinable way, the public mood acknowledges that murder, particularly deliberate murder, is a crime apart. Although not always treated as such, human life is sacred, and the survival of any civilized society worth the name will depend ultimately on respect for the sanctity of life. To take life away can be justified only in the most extreme circumstances and with the specific authority of the state. It is never open to one individual acting on his own account to terminate the natural existence of another human being. The extent to which human life, actual or potential, should be protected by the criminal law, especially in the difficult and uncertain areas of euthanasia, abortion, and suicide, is a profound question going well beyond the narrower focus in this chapter on the penalty for murder.[23]

[23] These issues are explored by Professor Glanville Williams in *The Sanctity of Life and the Criminal Law,* Faber and Faber, London, 1958.

The terrible finality of murder, its most distinctive quality, is mirrored in the offence of manslaughter. In most years, up to twice as many offenders are convicted of manslaughter as of murder. A substantial proportion of these, although a constant minority of the total, are found guilty of manslaughter on grounds of diminished responsibility, having originally been charged with murder. Each year there is a much smaller number of women, usually counted in single figures, who have killed their baby within twelve months of childbirth.[24] English law has long accepted that a woman's balance of mind may be affected by childbirth, and infanticide has been a separate crime since 1922.

If public opinion is generally tolerant towards the sometimes compassionate circumstances of a killing in which there is clear evidence of diminished responsibility or provocation, and even more so towards the occasional tragedies of infanticide,[25] it is verging on the indifferent in its attitude towards the largest category of all: the fatalities that result from the serious criminal offence of causing death by reckless driving. The fact that the slaughter on the roads has failed to arouse the same outright disapprobation as other forms of killing is no more than the most glaring example of the phenomenon that the careless, drunken, or reckless driver is not regarded by the public as a real criminal.[26] Hardly any of the stigma normally attached to crime rubs off on the offending motorist, who experiences little if any social ostracization, while juries are frequently more benevolently inclined towards motorists than other offenders. Whereas an arsonist who burns down a house with reckless disregard for the safety of the inhabitants is almost certain to go to prison (unless mentally disordered), the driver of a car who recklessly endangers other road users is not. That this should be so is understandable enough; there are countless motorists, and road transport is an essential feature of modern life. Those who judge

[24] The number of convictions of infanticide in each year since 1974 are as follows: 1975—5, 1976—6, 1977—6, 1978—8, 1979—7, 1980—10, 1981—7, 1982—6, 1983—11, 1984—2, 1985—8. *Criminal Statistics England and Wales 1985*, p. 59.

[25] On 14 March 1986 *The Times* reported that, in separate cases of infanticide two young women, one of whom was only 16 at the time, had been put on probation for killing their babies. In one case the probation order was for three years; in the other for two years, with a condition that psychiatric treatment was continued.

[26] See Barbara Wootton on motorists in *Crime and Penal Policy*, George Allen & Unwin, London, 1978, pp. 213–15.

may expect to be judged themselves one day. Moreover, virtually no driver involved in an accident goes out in his car intending to commit an offence, and typically all will regret bitterly the consequences once they have become involved in a breach of the law, particularly if it leads to personal injury or death. Even though the criminal conduct (the careless, reckless, or drunken driving) has a consequence that is accidental and unforeseen, nevertheless, questions of foresight and of natural consequences are well traversed ground in the law of homicide.

This is not a plea for harsher treatment of motoring offenders: it is simply to underline the significance of the connection between public attitudes and the sentencing of offenders. The law works in practice, it is generally accepted, and does not lead to injustice between one defendant and another. Since the element of calculation in advance is almost invariably absent, sentencing policy can have only an indirect deterrent effect. The penalties imposed by the courts in cases of this sort relate culpability and blameworthiness to the essentially accidental consequences of a driving offence. The retribution exacted will depend in the last resort on public expectations. Despite some recent poll findings indicating a hardening of public attitudes towards drink–drivers,[27] especially before Christmas, when the dangers are intensively publicized, it is probable that the tacit expectation that motoring offenders, other than in the most exceptional cases, merit comparatively lenient treatment will endure for a long time to come.

The finality and irreversibility of death applies to the convicted as well as to the victims. In many ways, the most conclusive argument against the death penalty is the possibility of error. However careful and thorough the groundwork, no human judgment can be infallible. Every year a few cases of wrongful conviction come to light, whether as a result of new evidence or of confessions by the true culprit or his associates. Where a man has been sentenced to a term of imprisonment he may be released by the courts or the Home Secretary, sometimes obtaining a pardon and occasionally being compensated financially. None of this can happen if he has been executed. If a

[27] A survey carried out by Gallup Poll inferred that drink driving was found to be more offensive to the general public than murder, drug addiction, or AIDS. *Daily Telegraph*, 12 Dec. 1986.

mistake should occur, it cannot be put right. When capital punishment existed, the greatest pains were taken by the courts and by the Home Secretary, in the exercise of the royal prerogative of mercy, to avoid the possibility of an innocent man being convicted and hanged. Yet according to Barbara Wootton, an authoritative commentator, in the thirteen years before the suspension of the death penalty in 1965, there were at least three cases in which serious doubts were raised as to whether the right man was convicted.[28] Timothy Evans was hanged for the alleged murder of his baby daughter, largely on the evidence of the notorious murderer, Christie. After two subsequent investigations it was officially admitted that Evans should not have been convicted of murder and executed. In two more cases, one in 1952 and the other in 1962, further facts became known after the supposed murderers' convictions which caused doubt to be thrown on the verdict. This is not to say that innocent men were necessarily hanged; just that, if a mistake was made, as the evidence strongly suggests was the case with Evans, it was irremediable.

In the debate on capital punishment, public attitudes are canvassed regularly by opinion pollsters. As usual, the questions asked are crucial, but the simple and direct question put by MORI goes to the heart of the controversy: 'Do you think the death penalty is ever justified or not?' This question has elicited consistently favourable responses. Since 1977 the highest proportion of those questioned who replied 'Yes, sometimes' was 81% and the lowest 74%. Supplementary questions were aimed at establishing which crimes were thought deserving to be punished by the death penalty. As in all opinion polls, it is the perceptions of respondents that are measured rather than truth or reality. But there seems no reason to doubt that to a majority of people the death penalty does mark an appropriate response for the most serious crimes of murder, although the poll data suggest that only minorities believe it should be applied to offences of rape, armed robbery, kidnapping, or the hijacking of aircraft. What emerges from the polls, backed up by much anecdotal evidence, is the clearest example to be found in the penal field where a majority in Parliament has stood out against

[28] L. Blom-Cooper (ed.), *The Hanging Question,* Duckworth, London, 1969, pp. 16–18.

the prevailing tide of popular opinion. The implications are more complex than they appear at first sight, and there is little that can be offered by way of conclusive proof that the incidence of murder is materially influenced by the penalty attached to the crime. Whether more knowledge and a wider understanding of the context in which murders occur would moderate the strength of popular feeling in support of a return to capital punishment is a moot point.

As it is, the issue of the death penalty is one that is best suited to parliamentary debate and decision, so enabling Parliament to discharge its true role of representing the nation in a sense broader than the purely numerical in resolving conflicts of opinion. The parliamentary system has many critics, but in this instance the House of Commons has shown that it is capable of exercising its independent judgment, determining decisively the direction of public policy, influencing opinion while at the same time responding to it.

V

So long as capital punishment existed, there was a justification for a mandatory penalty for murder. If murder was regarded as a crime apart, so was capital punishment regarded as a penalty apart. Thus, an equation was established and maintained between murder and capital punishment. In the great *History of English Criminal Law*, Radzinowicz and Hood record the attempts that were made in the nineteenth century to reform the law on murder and abolish the death penalty.[29] The abolitionist cause was making headway when it received a setback in 1868 from which it did not recover for generations. The place was the House of Commons and the speaker, John Stuart Mill. In a forceful and closely argued speech, we are told, he scornfully attacked abolition to such effect that his words were quoted for years to come. The rhetoric was undeniably memorable. In the gravest of cases, he declared, where there was 'no palliation of the guilt', where the crime was a consequence of the general character rather than an exception to it, it was entirely

[29] Sir L. Radzinowicz and R. G. Hood, *A History of English Criminal Law and its Administration from 1750*, vol. 5, Stevens, London, 1986, pp. 661–88.

appropriate 'solemnly to blot [the murderer] out from the fellow-ship of mankind and from the catalogue of the living'.[30] Mill did not consider the punishment of death as cruel; to him, it was merely hastening a man's inevitable demise. His speech won praise from *The Times* leader the next day for having 'the courage to disregard the sentimental and showily benevolent view of things'.[31] Thereafter, the momentum towards reform was checked and did not begin to build up again for well over half a century.

Although the equation between capital punishment and murder was finally severed in 1965, the legacy of the mandatory sentence remains to this day. The compromise offered to the retentionists was that a conviction for murder should auto-matically lead to imprisonment for life and that the courts should have no power to vary the penalty decided upon by Parliament. But life sentences, unlike capital punishment, were not unique; nor did they mean the offender would necessarily spend the rest of his natural life in custody. Ever since the first introduction of the life sentence as an alternative to transportation in the latter part of the nineteenth century, the Home Secretary has used his discretion to release lifers on licence after what he regarded as an appropriate period in prison had elapsed. An account of how this system works in practice is deferred until a later chapter. It is enough to make the point here that life imprisonment as a penalty does not stand in any unique relationship to murder since it is the identical penalty that the courts may impose in the most serious cases of homicide that fall short of murder. Other offences carrying a liability to life imprisonment include rape, armed robbery, wounding with intent to do grievous bodily harm, aggravated burglary, sexual intercourse with a girl under thirteen, and arson.[32]

The special quality of the life sentence is its indeterminacy: the period of time served may be longer or shorter than a fixed sentence. It is consequently wholly misleading to represent the mandatory life sentence as a watered-down version of the

[30] Ibid., p. 685.

[31] *The Times*, 22 Apr. 1868.

[32] There is a long list of offences carrying liability to life imprisonment, ranging from piracy and mutiny in the eighteenth century to hijacking aircraft and biological warfare in the twentieth century. See Appendix N to *Sentences of Imprisonment: A Review of Maximum Penalties*, HMSO, London, 1978, pp. 214–16.

previous mandatory penalty for murder. The death penalty was genuinely unique, and its application to murder could be argued on the lines advanced in Parliament from John Stuart Mill's day to the House of Commons debate in 1983. The convicted murderer who is executed cannot by definition offend again; nor do questions of release and future risk of offending arise. The life sentence, by contrast, is steadily being extended as the maximum penalty for a whole range of non-homicide offences. These may or may not be justified on other grounds, but each time a fresh extension is made it undermines still further the status of the life sentence as the symbolic penalty for murder.

All that remains in practice is a mandatory requirement to impose one specific type of penalty that is also generally available for use in other cases of serious crimes. But is life imprisonment the most suitable penalty for murder? A number (probably a not inconsiderable number) of High Court judges who try murder cases would prefer the courts to be given an unfettered discretion to impose either a life sentence, or a determinate sentence of imprisonment, or such other disposal as might be appropriate in the particular circumstances of the case. If the change were made, most of the difficulties described in the last chapter surrounding the defence of diminished responsibility and the psychiatric evidence given by expert witnesses would wither away. At present it is not open to the court to make a hospital order with restrictions where an offender is convicted of murder. For murder there is only one penalty: life imprisonment. The defence of insanity, in the very rare cases when it is used, of course, precludes conviction. Thus, the dividing line between murder and manslaughter, which in Section 2 Homicide Act cases means looking into the mind of the offender and reaching a judgment on his mental state and degree of responsibility, is of crucial importance. Even this distinction does not help with the borderline cases where a defendant has acted under great pressure, but is neither diagnosable as a psychiatric case for treatment nor excusable on grounds of provocation.

There are two possible lines of reform that can be pursued separately or together. The most straightforward approach, which was favoured by the Butler Committee on Mentally Abnormal Offenders as its first choice in advocating the abolition of diminished responsibility and the mandatory life sentence,[33] is to amend the law so as to make life imprisonment the maximum rather than the mandatory penalty for murder. As the Committee pointed out, cases of murder occur that can be adequately dealt with by a determinate sentence, whereas the use of the life sentence in cases that do not really call for it dilutes what should be the awe-inspiring nature of this penalty. Some manslaughters of a particularly brutal character may be more culpable than murders which may have mitigating circumstances. There are cases in which the mandatory life sentence is simply not appropriate, while in others it may be inhumane. To enable the trial judge to show humanity in the cases involving mental disorder, medical witnesses (according to the Butler Committee) 'sometimes have to stretch their conscience in testifying under Section 2 of the Homicide Act'.[34] In the most serious cases, the life sentence would continue to be used, possibly linked to a judicial recommendation of a minimum period of years to be served, if the court preferred that alternative to a lengthy determinate sentence for a fixed period of years. This is in fact a minimalist approach confined to the penalty for murder only, and leaving the offence of murder untouched. It has cogent arguments of principle as well as substantial practical advantages to commend it. The alternative is to go further and, taking a deep breath, abandon the highly unsatisfactory distinction between murder and manslaughter altogether. If the two were to be merged into a single criminal offence of unlawful homicide, defined by statute and carrying a maximum penalty of life imprisonment, then at last the clarity, accessibility, and consistency that have for so long been sought in this fundamental branch of the law would be within sight.

There is no shortage of expert opinion, since the penalty for murder has been considered by several specialist groups over the last few years including the Advisory Council on the Penal System (1978),[35] the Criminal Law Revision Committee (1980),[36] and the Parliamentary All-Party Penal Affairs Group

---

[33] *Report of the Committee on Mentally Abnormal Offenders* (Cmnd. 6244), HMSO, London, 1974, pp. 244–8.

[34] Ibid., p. 245.

[35] *Sentences of Imprisonment.*

[36] *Criminal Law Revision Committee 14th Report.*

(1985).[37] All have published criticisms of the mandatory sentence of life imprisonment for offenders convicted of murder, although the Criminal Law Revision Committee revealed that it was deeply, and almost evenly, divided on the issue and consequently made no recommendation but merely set out the arguments for and against the mandatory life sentence. We may note in passing that this split represented a shift of opinion, since the previous report in 1973 by the same committee on the subject of the penalty for murder had been unanimous in upholding the mandatory life sentence.[38] For whatever reason, none of these bodies addressed themselves to the seemingly more radical approach of a single offence of homicide: could it be for fear of walking out even further on the thin political ice?

VI

It is certainly true in this matter, perhaps more than any other in the penal field, that the politics of reform take precedence over the merits of reform. For the best part of the twenty years since abolition of capital punishment, the tactics of both the main parties in the House of Commons, backed up by official opinion in the Home Office which has advised successive Home Secretaries, has been to let sleeping dogs lie. It has been thought too risky to raise the floodgates in case the tide that would sweep into parliamentary channels would do more to destroy than to fertilize. The Butler Committee was forthright in stating that the evidence it had received had led it to conclude that 'many in the legal profession and other professions associated with the law believe in the abolition of the mandatory life sentence and in giving the courts the widest possible discretionary powers in sentencing. Our impression is that such a change is seldom advocated publicly because of the fear that it will be unlikely to commend itself to public opinion.'[39]

Would a single offence of unlawful homicide necessarily be

[37] *Report on Life-Sentence Prisoners,* Barry Rose, Chichester. The Committee reported in 1985, although the publication is undated.
[38] *Criminal Law Revision Committee 12th Report* (Cmnd. 5184), HMSO, London, 1973.
[39] *Report of the Committee on Mentally Abnormal Offenders,* p. 245.

even more controversial, a far shore that could not be reached because of the strength of parliamentary opposition and hostile currents of opinion outside? In 1985 the Law Commission published a model criminal code showing how the criminal law of England could be rationalized and codified.[40] The general provisions of the law and, by way of example, offences against the person were refined into a draft Bill of ninety clauses occupying (without the schedules) no more than thirty-eight pages. The measure was much more than a work of compression and restatement. The draftsmen, a team of lawyers from the Society of Public Teachers of Law under the chairmanship of Professor J. C. Smith, made it clear that, because of the gaps, inconsistencies, and arbitrary rules in existing law, the proposed codification represented a work of substantial reform. Their report was explicit: 'the code cannot reproduce inconsistencies. Where the inconsistency represents a conflict of policies, a choice has to be made to produce a coherent law . . .'[41] The report also criticized the lack of accessibility and uncertainties of the criminal law. Both murder and manslaughter were succinctly defined for the first time in statutory form, the latter being divided into voluntary and involuntary manslaughter. The defences of diminished responsibility, provocation, and excessive force were placed in a direct relationship to the statutory offences to which they would apply, and there were new clauses on mental disorder and incapacity. The Law Commission published the report of the criminal code team, together with the draft bill, as a document for discussion on which the views of the legal profession and the public were invited.

The attempt to codify the criminal law goes back to the mid-nineteenth century. Bills were introduced in Parliament in 1853 and 1878–82, by which time Sir James Fitzjames Stephen was the dominant driving force. As Home Secretary, Roy Jenkins came out strongly in favour of codification in 1967, speaking of:

too many archaic principles that have been handed down from precedent to precedent. As a result much of our criminal law is in many areas obscure, confused and uncertain. Yet no area of the law is of

---

[40] *Codification of the Criminal Law: A Report to the Law Commission*, Law Com. no. 143 (HC, 270), HMSO, London, 1985.
[41] Ibid., p. 19.

greater importance to the liberty of the individual, and nowhere is it more important that the law should be stated in clear and certain terms to take account of modern conditions.[42]

Since then, various aspects of criminal law have been subjected to detailed examination and proposals for reform have been advanced, some (although not all) of which found their way into the safe haven of the draft code. In a cautious introduction to the report containing the code, the Law Commission stated that codification remained the objective of its programme of reform for the criminal law and quoted the Lord Chancellor, Lord Hailsham, as saying: 'a good codification would save a great deal of anxiety, obscurity, consumption of judicial time, and so of costs'.[43]

If such authoritative opinion is willing to contemplate far-reaching legislation extending right across the full spectrum of the criminal law, why should it be thought that a more modest attempt to rationalize the law on homicide and the penalties attaching to it would be so objectionable? In their covering report, the team that drafted the code saw merit in this aim, stating unequivocally that 'the law on homicide could be enormously simplified by the abolition of the mandatory life sentence for murder and the merging of the crimes of murder and voluntary manslaughter'.[44] They added, however, that these were weighty matters of policy on which it was not for them to comment or make proposals. Echoes of Lord Butler's strictures can still be heard.

What contribution would the changes canvassed in this chapter make to countering crime, it may well be asked? Are they any more than legalistic logic chopping, all very well for urbane conversation in the Temple or the university common room, but remote from the harsh realities of violence in the streets, in the work-place, or in the home? What about particularly vulnerable victims and potential victims—young children and the elderly? Are they to be put at greater risk, or at less? It is hard to see how a departure from the mandatory life sentence would give heart to the calculating killer, since it is precisely the

---

[42] *Codification of the Criminal Law*, p. 3.

[43] Ibid., p. 12. Lord Hailsham's comments were made in an address to the Statute Law Society, 27 Oct. 1984.

[44] Ibid., p. 18.

deliberate murder planned in advance that would continue to attract a sentence of life imprisonment. Indeed, the natural consequence of a move towards reserving life sentences for the most serious offenders would be that the average period of time spent in custody would manifestly increase. This in turn would mean that life would be regarded as a sentence of greater severity. At present, the belief has spread that life 'only' means something between nine to eleven years' imprisonment. This calculation arises from a regular, although regularly fallacious, interpretation of the statistics published annually by the Parole Board. These show how long those life sentence prisoners who have been recommended for release by the Board in the course of the year will have been detained by the time of release.[45] The fallacy in calculating an average from the table showing the number of years served in relation to the number of prisoners recommended for release lies in the fact that such a calculation takes no account of those who have *not* been recommended for release: currently more than half the total considered each year.[46]

The second reason why any calculating criminal foolish enough to count on serving an 'average' life sentence may find himself gravely disappointed is that a large majority of the shorter terms of years in custody are served by prisoners whose crime was domestic in character and not planned in advance. These once-in-a-lifetime crimes are of course very serious, and the consequences for the victim are the same whether the killing was caused by a sudden loss of self-control or was premeditated. The requirements of retribution and deterrence to others, however, are served by a lesser period of imprisonment than, say, a bank robber who kills in the course of a raid that goes wrong, or an armed man who kills a police officer when he has been detected, or a terrorist who plants a bomb, or any of the other gravest forms of murder. In every case the circumstances of the crime, the background of the offender, and his subsequent progress in prison will be carefully assessed from the standpoint

---

[45] *Report of the Parole Board 1985* (HC, 428), HMSO, London, 1986, p. 10.

[46] In 1985, 344 cases of life sentence prisoners were considered by the Parole Board, of whom 76 were recommended for release and 200 as being not suitable for release. In ten cases the Home Secretary was unable to accept the Board's recommendation to release a prisoner on licence. *Report of the Parole Board 1985*, p. 10.

of the possible risk of his reoffending before any recommendation is made for release. These issues are pursued in a later
chapter, but the relevant point here is that, the longer the period
of time actually spent in custody by those sentenced to life
imprisonment, the greater the respect in which the life sentence
will be held. Life cannot mean life; it never has and it never will,
if it is applied indiscriminately to all those who are convicted of
what is one of the most varied of all crimes. Surely the thrust of
policy should be to make life sentences more forbidding as a
sanction. If the courts had the discretion to apply such penalties
as they thought fit to the less culpable murders, then the 'average'
period of time served by the more culpable offenders who were
sentenced to life imprisonment would rise sharply. Those who
put their trust in the deterrent effect of sanctions might be
expected to support this argument and ask whether the
continuous extension of the life sentence to more and more
offences is really wise if the result is to downgrade its credibility.

The year 1985 saw the life sentence extended as the maximum
penalty for attempted rape and trafficking in Class A drugs
(including heroin, LSD, and cocaine). In November 1986 the
Criminal Justice Bill included a clause increasing from fourteen
years to life the maximum penalty for carrying firearms during
the commission of crime. The maximum sentence for certain
firearms offences is already life imprisonment,[47] to which the Bill
proposes should be added possessing a firearm (or imitation
firearm) at the time of either committing or being arrested for
certain offences, and carrying a firearm (or imitation) with
criminal intent. Earlier in the year, when the proposal had been
forecast in a White Paper,[48] the Police Federation warned that it
would remove the differential between the punishment for
murder and the punishment for carrying a firearm during the
commission of a crime. The Federation's spokesman was
reported to have pointed out the danger that a criminal who was
trapped while carrying a gun would not face any deterrent from

---

[47] Possessing firearms with intent to endanger life, and using firearms with
intent to resist arrest, carry a liability to life imprisonment under the Firearms
Act 1968.

[48] *Criminal Justice: Plans for Legislation* (Cmnd. 9658), HMSO, London,
1986, p. 4.

using it to kill.[49] In his speech on the Bill's Second Reading, the Home Secretary rejected this criticism, declaring:

The proposal is for a maximum, not a mandatory sentence, and the court can differentiate. Other things being equal, the courts are bound to regard the use of a firearm as the more serious offence and to sentence accordingly. The effect of our proposal is that the highest penalties will be available for the worst cases. Far from weakening the deterrent against pulling the trigger, the proposal is a strong incentive for the firearm not to be there in the first place.[50]

There can be no certainty in the way sanctions will bite on those who are most likely to be affected by them. As the two statements cited above demonstrate, opinions may differ even among the public authorities who are closest to organized crime and serious offending. It is important for political responses of this sort to be scrutinized with particular care if they are not to lead to a situation in which the efficacy of penal sanctions is weakened rather than strengthened. In practice, as the Home Secretary recognized, greater confidence in the process of criminal justice, so fundamental an aim, is more likely to be enhanced by the actual effects of the sentences imposed by the courts, and the time served in custody, than by increasing maximum sentences as an expression of public concern. 'Tough reform aims to make criminal pay' was *The Times* front-page headline on the White Paper in March 1986. The danger is that the emphasis on presentation is so compelling that it can mask the relevance of a discernible trend, seemingly gathering pace, to extend the life sentence to an ever-widening range of non-homicide offences. Proclamation and denunciation have their place in sentencing policy, but it would be regrettable if the price paid was to undermine the significance of the life sentence as the symbolic penalty attached to the ultimate criminal offence: taking the life of another human being.

[49] *The Times,* 7 Mar. 1986.
[50] *Parl. Debates,* HC, 106 (6th ser.), col. 469, 27 Nov. 1986.

6

Youth, Delinquency, and Drugs

I

One of the most striking of all criminal statistics, which never fails to be greeted with expressions of astonishment and alarm whenever it is quoted, is the evidence that, of all males born in England and Wales in 1953, nearly one-third had been convicted of an indictable criminal offence by their twenty-eighth birthday.[1] More than half of these had been convicted only once, while about 15% (representing 5.5% of the total sample) between them had accumulated 70% of all the crimes committed by members of this age group and dealt with by the courts.[2] Note the last words: 'dealt with by the courts'. Many crimes are not reported at all and so never come to the notice of the police. Others will have been reported to the police, but the culprits may not have been identified or the evidence may have been inadequate to bring a prosecution. In yet other cases, especially those of juveniles, an official police caution may have resulted. None of this offending is included in the statistics referred to above, nor are numerous minor or motoring cases which are dealt with in the Magistrates' courts. These exclusions, so extensive in volume, highlight the significance of the research findings still further.

There are, as always, qualifications. The research is continuing, and there are signs of some more encouraging trends on the way. Disturbing though it is to learn that a proportion as high as nearly one-third of young males can expect to be

[1] Details are given in 'Criminal Careers of Those Born in 1953, 1958 and 1963', *Home Office Statistical Bulletin* 7/85. This bulletin analyses convictions at all courts for 'standard list' offences, which are broadly the equivalent of indictable offences and exclude some minor summary and motoring offences dealt with at Magistrates' courts.
[2] Central Statistical Office, *Social Trends 16*, HMSO, London, 1986, Table 12.12, p. 189.

convicted of an indictable offence by the age of twenty-eight, some consolation can be found in the fact that a majority of these young men do not re-offend. For many of them, their single conviction stands out as an inglorious pinnacle in an adolescent phase of delinquency centred around property offences, most of them relatively minor, which may be brought to a conclusion when they come to face the unpleasant consequences of their offending in a court setting. For a minority, however, court proceedings have little deterrent effect, anti-social habits being too deeply engrained, or the magnetic attraction of the peer group too irresistible, and the pattern of persistent offending emerges. The research shows that those convicted of burglary are more likely to be reconvicted, whatever the sentence—custodial or otherwise—imposed by the court, than those convicted of other types of offence.[3] Most sombre of all is the finding that the peak age for convictions for indictable offences of males born in 1953, 1958, or 1963 was seventeen. If cautions are included, the peak age for offending (i.e., recorded offences) falls to nearer fifteen for males and fourteen for females.[4]

There is a marked difference between the sexes. For girls and young women, the proportionate number of offenders represented in the age group is much less: only about 6% of those born in 1953 incurred one or more convictions for indictable offences by the age of twenty-eight. In the same way as the men, it is a very small proportion of females born in each year who account for most of the offending committed by women.[5] Over the last two decades, however, the rate of known offending by females has risen considerably faster than the comparative rate for males.[6] This is one move towards equality that is manifestly in the wrong direction.

The encouraging trends are demographic, and the horizon is getting closer. If crime is concentrated disproportionately among young people, mainly male, as is the case, it follows that the size of the age group under twenty-eight as well as its relationship to the population as a whole, is highly relevant. The postwar bulge in the birth rate transformed the demographic and social composition of British society. In the United Kingdom there are

[3] *Home Office Statistical Bulletin* 7/85, p. 2.   [4] Ibid., p. 4.
[5] Ibid., p. 1.   [6] *Social Trends 16*, p. 189.

now half as many again young men aged between seventeen and twenty as before the Second World War. In 1955 there were approximately 1 million young males in that age group: by 1983, when the size of this group in the general population peaked, it had risen to about 1.7 million.[7] It is tempting to generalize about bad housing and social alienation, to dwell on the restricted opportunities for educational development and employment, and to shake our heads over racial discrimination. We do not know if any of these factors cause crime in themselves, but they seem to provide the fertile soil in which delinquency takes root. Parental control and example in the home is one of the surest determinants of constructive attitudes and behaviour,[8] but it is an influence that has been in retreat when faced with the pervasive ethic of self-expression.

In the United States the pattern has been the same: a postwar baby boom leading to the rapid expansion of the most crime-prone age group. America has experienced more of a collapse than an evolutionary change in its popular culture, the impetus being accelerated by the twin forces of racial tensions, leading to the civil rights movement, and the Vietnam War and its aftermath. As acceptance of the moral legitimacy of the institutions embodying the American way of life has declined, so offending has risen. In the same way as in Britain, the higher proportion of young men in the population at large has led to an increase in the volume of crime, but pro rata, the rate of offending by young males of a given age (the age-specific crime rate) has also gone up. In one study, quoted by James Q. Wilson of Harvard in his notable book, *Thinking About Crime*,[9] it was estimated that a delinquent boy born in Philadelphia in 1958 was five times more likely to commit a robbery than one born in the same city in 1945.[10]

---

[7] Sir B. Cubbon, *Crime—Our Responsibility*, James Smart Lecture, London, 1985, p. 4.

[8] Research findings indicate that the nature of the supervision exercised by parents is an important factor in delinquency. See D. Riley and M. Shaw, *Parental Supervision and Juvenile Delinquency* (Home Office Research Study no. 83), HMSO, London, 1985.

[9] J. Q. Wilson, *Thinking About Crime* (rev. edn) (Basic Books, New York, 1983), contains a fine chapter on 'Crime and American Culture' at pp. 223–49. James Q. Wilson is Henry Lee Shattuck Professor of Government at Harvard University.

[10] Ibid., pp. 223, 279 n.

After a period of steady growth in the 1970s recorded crime in the United States decreased during the early 1980s. Academic commentators were inclined to attribute the decline to the falling birth rate of earlier years, with politicians giving more weight to heavier sentencing. In 1985, however, the indications were that the figures had begun to rise again.[11] It would be unwise to pin too much hope on a really significant change in the tide of overall offending as the birth rate falls. There are, none the less, some heartening straws in the wind. Home Office statistics published in 1985 indicate a reduction of 16% in the number of young offenders in the age group ten to fourteen years sentenced for indictable offences. Contributory factors were a decline in the birth rate in the early 1970s and the greater use of cautioning instead of prosecution.[12] Forecasts of the impact of population changes on crime are guarded, but the predictions point towards a fall of the order of 1% per annum arising from this factor over the next ten years, as opposed to an annual increase of about the same amount during the previous two decades. Demographic projections suggest there may be some 25% fewer adolescents in the mid-1990s than in the peak year of 1983.

II

Special measures to deal with offending by juveniles have long been a feature of the English penal tradition. Although for centuries no distinction was made between children and adults committed to gaol, by the late eighteenth and early part of the nineteenth centuries some allowances began to be made for the age and immaturity of children and young offenders. Today, with few exceptions, juvenile defendants between the ages of ten (the age of criminal responsibility) and seventeen are tried in separate courts from adults. The origins of the juvenile court date back to the opening years of the present century, but the emphasis on correction and the improvement of youthful character well preceded the objective of rehabilitation for adult offenders. As

[11] On the basis of the Uniform Crime Report (UCR) figures published annually by the US Department of Justice, Washington, DC.
[12] *Home Office Statistical Bulletin* 12/85 and comment in *Criminal Law Review* (1985), p. 413.

early as 1788, a school for the reform of young criminals was established in London. Parkhurst, the grim maximum security prison on the Isle of Wight, followed in 1838 as a place of detention and correction for juvenile offenders, with a regime that combined industrial training with religious and educational instruction.[13] By the mid-nineteenth century the Victorians had made wider provision, with reformatory schools for young offenders and industrial schools for training children in need of care and protection. Borstal training, a radical and pioneering concept in its day, was adopted in 1908, although the first experiments dated from 1900.[14] The credit for giving statutory form to both Borstal institutions and the probation service lies with Herbert Gladstone, son of W. E., who was Home Secretary from 1905 to 1909.[15] But much of the inspiration for these and other far-reaching reforms came from a renowned penal reformer, Sir Evelyn Ruggles-Brise, who served as Chairman of the Prison Commissioners for over a quarter of a century from 1895 to 1921.[16]

It was, however, Gladstone's more famous successor, Winston Churchill, mindful of his own captivity as a prisoner of the Boers, who left a more conspicuous mark on penal policy by humanizing and improving prison conditions and forcing through changes which made it possible for vastly fewer offenders to be sent to prison.[17] This was a great achievement, never since rivalled, and it was preceded by an eclectic process

[13] D. M. Walker, *The Oxford Companion to Law*, Clarendon Press, Oxford, 1980, p. 697.

[14] The origins of the Borstal experiment are described by Sir L. Radzinowicz and R. G. Hood in *A History of English Criminal Law and its Administration from 1750*, vol. 5, Stevens, London, 1986, pp. 384–7.

[15] Created a peer and appointed as the first Governor-General of the Union of South Africa in 1910, Gladstone complained that Churchill had taken the credit for his schemes of penal reform. See R. S. Churchill, *Winston S. Churchill*, vol. II, *Young Statesman 1901–1914*, Heinemann, London, 1967, pp. 386–7.

[16] For a biographical assessment of Ruggles-Brise's contribution to penal policy and practice, see Radzinowicz and Hood, *A History of English Criminal Law*, vol. 5, pp. 596–9.

[17] In the year in which Churchill became Home Secretary, 1908/9, imprisonment resulted from approximately 184,000 cases before the courts, more than half being committals in default of payment of fines and a third for drunkenness. By allowing time to pay fines and restricting imprisonment to more serious offences, the numbers imprisoned were reduced over a ten-year period from 95,686 for non-payment of fines in 1908/9 to 5,264 in 1918/19, and from 62,822 for drunkenness in 1908/9 to 1,670 in 1918/19. Ibid., pp. 387–9.

of consultation in which Churchill showed himself willing to receive advice from such varied correspondents as John Galsworthy, W. T. Stead, and Wilfred Scawen Blunt, the last two of whom had also been imprisoned in their time. Some of his ideas provoked head-shaking in the Home Office, and 'Homeric laughter' was said to have been the response of Ruggles-Brise to at least one scheme.[18] But the Head of the Department, Sir Edward Troup, noted magnanimously: 'Once a week or oftener, Mr Churchill came to the office bringing with him some adventurous or impossible projects; but after half an hour's discussion, something was evolved which was still adventurous but not impossible.'[19] After seven months, the new Home Secretary was ready to put forward his prison reforms. Churchill's proposals for 'a scientific and benevolent measure, dealing with prisons and the punishment of offenders' were set out in a lengthy personal letter to Asquith dated 26 September 1910. The text was later incorporated almost verbatim in a Cabinet Paper titled 'Abatement of Imprisonment' and is published in the companion volume to the official *Life*.[20]

Churchill was also interested in young offenders and introduced legislation to ensure, *inter alia*, that no young person between the ages of sixteen and twenty-one should be sent to prison unless he was incorrigible or had committed a serious offence; that if a prison sentence was found necessary it must be for more than one month (many young people were serving very short terms of imprisonment, sometimes only a few days for trivial offences); and that, when justified, sentences of imprisonment were to be of a positively curative or educative character and not merely punitive. These reforms produced results only gradually, but within a decade the picture had been transformed. Whereas in 1910 there were 12,376 boys and 1,189 girls under twenty-one sent to prison, by 1919 the numbers had fallen to 3,474 and 762, respectively.[21] The massive social dislocation caused by the First World War, and the loss of so many young

[18] Ibid., p. 392.
[19] Radzinowicz and Hood, *A History of English Criminal Law*, vol. 5, pp. 774–5.
[20] R. S. Churchill, *Winston S. Churchill*, Companion vol. II, pt 2, *1907–1911*, Heinemann, London, 1969, pp. 1198–1203.
[21] Churchill, *Winston S. Churchill*, vol. II, *Young Statesman*, p. 390.

men, may have been contributory factors to the decline accelerated by changed policies.

In 1933 the distinction between reformatory and industrial schools was abandoned, the resulting institutions being renamed 'approved schools'. Many of these approved schools have subsequently been closed, and the remainder have become absorbed into the system of community homes. Borstal training, a sentence of indeterminate custody imposed by the court on selected young offenders between the ages of sixteen and twenty-one, originally brought with it an idealistic approach towards the reform of trainees by character-building and example. By the time it was abolished, being replaced by fixed terms of youth custody in the Criminal Justice Act 1982, disenchantment had set in, and few professionals felt that it was any longer the best way to treat young people in custody. The 1982 Act reasserted Churchill's determination that the court should not pass any form of custodial sentence unless it was satisfied that no other way of dealing with a person under twenty-one was appropriate,[22] but it seems unlikely that in the conditions of today the practical results will be any more than a faint echo of the past.

Institutions for young people appearing before the courts have proliferated. In addition to the ex-Borstals and young offender prisons used for those serving youth custody sentences, there are remand homes for observation and assessment, remand centres for those remanded in custody and awaiting trial, detention centres for short periods of custody not exceeding four months, and youth treatment centres for young persons convicted of very serious crimes (including homicide) and sentenced under Section 53 of the Children and Young Persons Act 1933. The youth treatment centres also contain some of the most difficult young people who have not been convicted of a crime, but who are subject to care orders and require a higher degree of security and supervision than local authority social service departments

---

[22] Section 1(4) of the Criminal Justice Act 1982 specifies that, where a person under 21 years of age is convicted or found guilty of the offence with which he is charged, the court may not make a detention centre order, or pass a sentence of youth custody or custody for life, unless it is of the opinion that no other method of dealing with him is appropriate because it appears to the court that he is unable or unwilling to respond to non-custodial penalties, or because a custodial penalty is necessary for the protection of the public, or because the offence was so serious that a non-custodial sentence cannot be justified.

can provide. Non-custodial sanctions available to the court include probation or supervision orders, supervised attendance centres, fines and compensation orders, absolute or conditional discharges, and above all the potentially constructive and not simply punitive community service orders. Community service has grown rapidly, from over 13,000 orders in 1979 to 33,800 in 1985. The scheme has been extended throughout England and Wales and is now available for all offenders aged sixteen and over.

Between them, these sanctions represent the formal, institutionalized channels through which society's response to the offences committed by young people is expressed. It can no longer be claimed that the range of alternatives to imprisonment is too limited; indeed, further additions in pursuance of the age-old search for panaceas are likely to lead to more confusion and even less precision in sentencing than applies at present. It is disappointing that the so-called alternatives to custody seem to be used by sentencers as much as alternatives to each other as alternatives to custody. If measured against current levels of crime, and the rate of increase over recent years which may now be beginning to taper off (although it would be premature to make any bold assertions), there is little to suggest that penal sanctions imposed after the event have been an effective counter to offending by young people. Incarceration may prevent the commission of crimes during the time an offender is in custody, but the price may be a deepening of criminality and so a heightening of risk for the future.

For this reason, the work done by NACRO and other voluntary organizations with young ex-offenders, some of it within the ambit of the Youth Training Scheme (YTS) and supported by public funds, is of the first importance. A substantial minority of young people who are eligible for YTS are reported as having come from backgrounds with social and personal problems which, to say the least, leave them ill-equipped to take advantage of training opportunities and to compete in the labour market on an equal footing with others.[23] In 1985 an estimated 18% of the boys and 3% of the girls eligible for YTS had convictions for criminal offences (excluding

[23] I. Crow and P. Richardson, *Youth Training and Young Offenders*, (Research and Development no. 24), Manpower Services Commission, London, 1985, p. 1.

motoring), and most of these young people had the further handicap of low educational standards as well as backgrounds of social or material deprivation or both. In 1984 approximately 2,400 young ex-offenders, or those thought to be at risk of offending, were taking part in local YTS training schemes pioneered and managed by NACRO. These provided opportunities for work experience and training, either with employers or in special workshops or projects. About a quarter of the trainees had served a custodial sentence, most commonly in a detention centre, while as many as two-thirds were subject to some sort of statutory supervision, either by the probation service or the local authority social services departments, at the time of joining a scheme. Others had court cases pending.[24] Research commissioned by the Manpower Services Commission indicated a mixed reaction by trainees, but a worthwhile number of placements was arranged and at least some young people lasted the course and were successful in obtaining a job as a result of a placement. Of the rest, about a quarter stopped attending of their own accord before completing the programme, with no known alternative to go to, and one in nine departed involuntarily to a penal institution.[25] Disappointing though the last statistic is, the likelihood is that the proportion returning to detention centre or youth custody would have been considerably higher without the stimulus and opportunities provided by the NACRO schemes within the YTS programme.

Community Service Volunteers (CSV), another well regarded national voluntary organization, although one that is not concerned primarily with offenders, enables between 200 and 300 young offenders each year to spend a few weeks at the end of a sentence of youth custody helping elderly, handicapped, or otherwise disadvantaged people in the community. The scheme aims to encourage personal development and self-reliance, the volunteers working under a CSV project supervisor and face to face with those who need their help. It has been in operation since 1971, and in 1985/6 225 young people serving sentences in fourteen closed youth custody centres (YCCs) worked as CSVs, usually for a period of a month, although sometimes for longer. With the agreement of the Home Office, those who

---

[24] Crow and Richardson, op. cit., pp. 5–6.   [25] Ibid., p. 7.

volunteer, and are assessed at the YCC as being suitable, are released under special arrangements before their normal release date to perform a specific task of community service. Out of the 1985/6 total of over 200, only eight failed to complete their month and were recalled to custody. Of these, two had re-offended while on their placement.

These are just two examples of programmes aimed at preventing the further commission of crime by young people by holding out the prospect of a better and more satisfying way of life. Of course, by no means all young offenders have the capacity to respond positively, and out of those who are selected a number will fall by the wayside. But prevention before, rather than punishment after, is the way ahead, and it deserves more attention and priority. I return to this theme in the later chapters. It is also an approach that fits well with the increasing emphasis that is being put upon crime prevention measures, initiated and carried through by citizens, individually and in groups, as well as by agencies of the state, so as to pre-empt or restrict the opportunities for crime.

Although thought-provoking, the overlap between drug-taking and offending by young people which results in appearances in court should not be exaggerated. Alcohol and solvent abuse pose similar problems, the severity of which tends to vary with the age ranges. Nevertheless, it is a fact that large numbers of those who make their way into young offender establishments are found on arrival to have been using drugs, although relatively few will have attracted convictions for possession or trafficking. Some others will have committed crimes in order to raise funds to finance their drug habit; but a majority will have been sentenced for what, on the face of it at least, are non-drug offences. Prison staffs now rate the drug problem as a serious threat to the stability of their establishments, partly because an inmate who has been taking drugs is likely to be less susceptible to the normal methods of control and discipline. This can take two forms: 'flashbacks', when an inmate behaves in a bizarre way, as if reliving some of his experiences under drugs; and the problems of drugs being smuggled into establishments, which typically occurs as a result of such necessary, but sensitive, facilities as visits, gift parcels, home leaves, and temporary releases. Rarer, and less obvious, are drug-related behavioural changes, sometimes reversible, such as personality disturbances and aggression, which may have contributed to a whole range of criminal acts including violence to the person as well as property offences.

It is not necessary to seek proven causal connections before contemplating the deeper perspectives that unfold in the relationship between offending and drug abuse. Is there anything in our contemporary society, in the way that private lives of individuals are integrated into an interlocking series of public communities, that has led to this terrible breakdown in relationships? We cannot allow ourselves to accept as inevitable so injurious a flaw as one generation succeeds another. We must strive to identify the root causes of offending by young people, and to clear our minds about what are the most promising avenues for intervention and action.

III

Parole Board members are well placed to obtain an insight into the criminal careers of young people from the detailed reports that are submitted on the large number of offenders under the age of twenty-one who are eligible to be considered for parole each year.[26] Scrutiny of these often revealing case histories leaves a powerful impression that in a majority of cases (although not all) the problems of deviancy start in the home, if there is one, and at an early age. Very young children who are abandoned and taken into care, or are looked after by relatives, or who remain in the family home to be neglected or abused, are put at risk of growing up as the criminals of tomorrow. Harrowing stories abound of one or both parents who are unable to cope adequately with the responsibility of rearing a family, or

---

[26] The Criminal Justice Act 1982, which came into effect on 24 May 1983, introduced a number of changes in the treatment of young offenders. Borstal training and sentences of imprisonment for offenders under 21 were abolished, being replaced by determinate sentences of youth custody. In 1985, 5,991 young offenders were eligible for parole, of whom 4,217 (70.4%) were recommended for parole. Not all were reviewed by the Parole Board, since substantial numbers of shorter sentenced prisoners are considered only by local review committees. See *Report of the Parole Board 1985* (HC, 428), HMSO, London, 1986, p. 9.

who are mentally impaired or disordered, or addicted to alcohol or drugs. Violence in the home, whatever the cause, appears to be distressingly commonplace. Aggression, verbal as well as physical, resentment or frustration may be vented by one adult partner against the other, or against a vulnerable old person in the household, or, most poignant of all, towards a child. Even where the violence and aggression is confined to adults, it cannot fail to have an unsettling effect on the children who witness it. Sexual abuse within the home, for so long a taboo subject, can result in profound and lasting damage to child victims. It is remarkable how many convicted adult sex offenders were themselves subjected to sexual assaults earlier in their lives. Nor is it exceptional to find references to physical injuries, including those to the head, whether accidental or non-accidental, suffered in childhood. These may cause brain damage and subsequent epilepsy, which is much commoner than expected in the prison population.[27] Partly for this reason, in the case of serious crimes of violence a neurological or neuro-psychiatric examination, including EEGs and computerized brain scans, often takes place to check for the presence of any brain injuries or malfunction.[28]

Time and time again, the even language of social enquiry reports reveals a pattern of emotional deprivation that runs parallel with and is reinforced by material deprivation. Children in large families clamour for parental attention in overcrowded and inadequate accommodation. Parents quarrel and separate, new partners come and go. Money is short and often badly managed. Bills go unpaid, debts mount up, essential repairs are not carried out. Relations deteriorate with landlords and local authority housing departments. Representatives of outside agencies get drawn in: child care, housing, educational welfare, health visitors, and probation officers. Most are well intentioned, and all are hard pressed for time and money. Sometimes a social

---

[27] J. C. Gunn, cited by T. A. Betts, 'Epilepsy and Psychiatry', in J. Laidlaw and A. Richens (eds.), *A Textbook of Epilepsy* (2nd edn), Churchill Livingstone, London, 1982. Since 1978, John Gunn has been Professor of Forensic Psychiatry at the Institute of Psychiatry, University of London.

[28] The EEG is a test of the electrical signals arising from the brain. They are often abnormal in the presence of epilepsy or other disorders. The computerized brain scan (CT scan) is a form of X-ray which detects alterations in brain structure, for example as a result of scarring due to birth injury, or head injury caused by a road traffic accident or battering.

worker or other visitor can become a rock of support and guidance for a family in trouble, but more often their periodic intervention (invited or uninvited) is regarded with a mixture of suspicion and passive acquiescence.

Outside the home, as the child grows up, there is the pervasive and not always benevolent influence of the peer group. Much has been written about the nature of delinquency and the sociology of delinquent adolescent groups.[29] Rejection within the family, poor achievements at school, and low self-esteem can bring youths together into sub-cultures with values of their own. The violence, destructiveness, and vandalism that are so hard for an outsider to understand can be interpreted as a defiant reaction against respectable society where such things are not done. The status hitherto denied in the home and school can be earned by a different route in a delinquent world, with its own customs and standards, where such things are done.

Children who are likely to become delinquent, and the extent to which delinquency can be predicted from knowledge about personality, family, and social background—questions that are wholly relevant for policy-makers—have been systematically investigated by the Cambridge Study in Delinquent Development over more than two decades. A sample of some 400 boys aged eight was originally recruited in the early 1960s. All were living in a working-class area in south London where there were no private schools and where the bulk of the housing belonged to the local authority. As much information as possible was gathered about the young boys and their families by the time they reached the age of ten and before the onset of any court appearances. The sample was then followed up and reinterviewed over a period of more than twenty years. By the end of the first decade, at the age of eighteen, one quarter of the sample had incurred convictions for criminal offences. Further interviews of the same group at the age of twenty-five showed that the proportion that had acquired a criminal conviction had risen to about one-third, thus reflecting the findings reported in *Social Trends*.

[29] See A. K. Cohen, *Delinquent Boys* (1956), and D. M. Downes, *The Delinquent Solution* (1966), both published as part of the International Library of Sociology and Social Reconstruction (founded by Karl Mannheim) by Routledge & Kegan Paul, London.

The Cambridge research, which is still continuing and is the most comprehensive of its kind ever carried out in this country, has enabled a delinquent minority to be compared at regular intervals with a similar but non-delinquent group. The findings have been published in four titles in the *Cambridge Studies in Criminology* and cannot easily be summarized.[30] In general, the research bears out the proposition that those who came before the courts had family backgrounds and personal characteristics that were notably less favourable than non-delinquents from the same areas. While the survey does not suggest that individuals can be identified at an early age as potential offenders with a high degree of accuracy, it does throw light on those who are most at risk and why.

High on the list of significant precursors of delinquency is separation caused by a breakdown in the marriage of a young person's parents. The Cambridge Study, endorsed by other research, validates the broken home as a breeding ground for subsequent offending, finding that boys whose parents were deserted, separated, or divorced included a high proportion of delinquents. That the crucial factor was marital breakdown, and the discord preceding separation, rather than the actual absence of one parent from the home, was supported by the evidence that loss of a parent through death from natural causes or accident, or temporary absence through illness, bore relatively little relationship to offending. Those adults who are quick to condemn the delinquent tendencies of young people should pause to note the strong influence of parental behaviour on juvenile offending.

Following closely behind the home environment and the quality of parental care and responsibility as factors relating to delinquency are the immediate neighbourhood and the school. The imposition of harsh or erratic discipline in the home, with little or no emphasis on rewards or encouragement for achievement, may be mitigated by a school that places due emphasis on

---

[30] Cambridge Studies in Criminology, vol. XXV: D. J. West, *Present Conduct and Future Delinquency*, 1969; vol. XXXIV: D. J. West and D. P. Farrington, *Who Becomes Delinquent?*, 1973; VOL. XXXV: D. J. West and D. P. Farrington, *The Delinquent Way of Life*, 1977; vol. L: D. J. West, *Delinquency: Its Roots, Careers and Prospects*, 1982, all published by Heinemann Educational Books, London. The General Editor of the series is Sir Leon Radzinowicz.

educational attainment and application, linking praise with sanctions, and ensuring that disciplinary measures are applied fairly and consistently throughout; the practice being understood by both teachers and pupils. Conversely, a school with low educational standards and uninspired teaching, one that is indecisive and inconsistent in its attitudes towards discipline and rewards, can expect more than its fair share of unruly and troublesome pupils with a higher-than-average disposition towards truancy. While school delinquency rates generally reflect the inbred characteristics of the original intake, it would appear that some schools do have the effect of promoting delinquency.[31]

In most cases the school and the home will be located in the same neighbourhood; and, as every probation officer working in an inner-city area knows, some neighbourhoods are far more inclined towards crime and delinquency than others. It is hard to say with any certainty why some local authority housing estates have become so prone to crime, but the concentration of large numbers of people in relatively confined spaces and the method by which the housing is allocated are likely to have something to do with it. The design and construction of housing in such a way as to reduce the opportunities for crime and vandalism is now receiving more attention, and I return to this important matter in a later chapter on crime prevention. With limited facilities for legitimate recreation and leisure, boredom and aimlessness, the two best recruiting sergeants for the delinquent gangs, have full reign. Socially acceptable outlets are needed for what Lord Scarman, in his report on the Brixton riots, called 'the exuberance of youth', if young people are not to become frustrated and diverted towards criminal ends.[32] High youth unemployment adds enforced idleness to boredom for many school-leavers, despite the opportunities offered by the Youth Training Scheme. The drinking club, the amusement arcade, the all-night disco, and the street corner become the focal points of a sub- or anti-culture; magnets for the disaffected and the delinquent in neighbourhoods that bristle with countless opportunities for crime. It is there, too, that drugs change hands.

---

[31] West, *Delinquency,* p. 99.
[32] *The Brixton Disorders, 10–12 April 1981. Report of an Inquiry by the Rt Hon. Lord Scarman* (Cmnd. 8427), HMSO, London, 1981. Reprinted 1986.

IV

It is a misconception to think of the abuse of dangerous drugs as the prerogative of the young or the rootless. The illegal import and distribution networks are professionally organized, the profits are enormous, and reference to the implications of organized drug trafficking was made in Chapter 3. Then too, there are many adult addicts—the lonely, the insecure, and those seeking to escape from their problems, whose lives and careers have been blighted by the misuse of drugs. While younger addicts often let drug-taking disrupt all aspects of their lives, older addicts may struggle to keep a normal life going despite the need for drugs, or sometimes with the aid of them.

The statistics relating only to the very tip of the iceberg— addicts notified to the Home Office—reveal some interesting comparisons when analysed by age as in Tables 9 and 10. It should be emphasized that these statistics represent a small proportion of chronic misusers, notably those addicted to heroin, or its derivatives or substitutes, and cocaine: LSD, amphetamines, and cannabis are excluded from the figures. Nor can notification, a statutory obligation placed upon doctors attending patients who are addicted to specified drugs,[33] take account of the much larger number who have not sought medical treatment and so will not have come to notice. The figures in Table 9 refer exclusively to those coming to notice for the first time.

The full extent of drug-taking and addiction in Britain is impossible to quantify, but it seems unlikely that the number of regular users of notifiable drugs in 1985 (of whom over 90% are addicted to heroin or other opiates[34] or synthetic substitutes) will have been less than 80,000 people. The total number of heroin addicts known to the Home Office in 1985 was 8,832, and the

---

[33] The Misuse of Drugs (Notification of the Supply to Addicts) Regulations 1973 require any doctor to notify the Home Office in writing within seven days if he attends a patient whom he considers to be, or has reasonable grounds to suspect is, addicted to any of a list of named controlled drugs. These include heroin and other opiates, or synthetic substitutes, and cocaine.

[34] The term 'opiate' means any drug derived from the opium poppy, while 'opioid' refers to synthetic substitutes. Confusion results when the description 'opioid' is used to include natural opiates.

TABLE 9. *Drug addicts notified to the Home Office: those coming to notice for the first time, by age group, 1974–85*

	Under 21	21–24	25–30	all ages
1974	281	310	146	870
1975	192	351	232	922
1976	162	354	306	984
1977	184	353	305	1,109
1978	207	432	420	1,347
1979	206	486	516	1,597
1980	257	459	504	1,600
1981	357	644	697	2,248
1982	489	768	826	2,793
1983	879	1,150	1,081	4,186
1984	1,206	1,544	1,363	5,415
1985	1,531	1,932	1,459	6,409

TABLE 10. *Proportion of newly notified drug addicts aged under 25 and under 30, 1974–85*

	% under 25	% under 30	Comment
1974	68	85	Many younger addicts
1977	48	75	Broader age range, rising numbers
1980	44	76	Broader age range, rising numbers
1983	48	74	Broader age range, rising numbers
1984	51	76	Trend to under-25s re-emerging
1985	54	77	Trend to under-25s re-emerging

entire addict population is now believed to be about ten times greater. If occasional use is added, the figure will be even higher. In Table 11 some qualitative differences between younger and older addicts are illustrated.[35]

Two characteristics in particular distinguish the adolescent from the mature drug-taker: the desire to experiment, and the pressure to conform with a peer group. When joined with the

[35] I am indebted to Dr George Birdwood for the preparation of these three tables. Tables 9 and 10 are derived from the annual statistics of drug addicts notified to the Home Office, while Table 11 represents a subjective assessment based on clinical experience.

TABLE 11. *Qualitative differences between younger and older drug addicts*

Feature	Younger (teens–30)	Older (30 & over)
Drug		
Types	More	Fewer
Dosage	More	Less
Pattern	More erratic	Less erratic
Sources	More illicit	Fewer illicit
Behaviour	More erratic	Less erratic
Criminal acts		
Drug thefts	More	Fewer
Money-raising	More	Less
Other crime	More	Less
Illicit supply	More involvement	Less
Compulsive drug-seeking	Very marked	More controlled
Introducing others to habit	More	Relatively little
Group activities	More	Often solitary
Ill effects on: Health Nutrition Work Behaviour Relationships	More	Often less severe
Risk to life	Greater	Less
Acceptance of therapy or advice	Only for immediate problem (overall problems denied)	More long-term (overall problems admitted)
Prospects of cure	Small	Somewhat greater

addictive quality of dangerous drugs, these are the prime causes of one of the most insidious social evils of the day, in Britain as in other advanced industrial societies. Young people have a natural curiosity and want to experiment. They hear a lot of talk about drugs from their friends and acquaintances, coinciding with recurring and sometimes sensational media coverage which can encourage the belief that taking drugs is far more widespread than it actually is. It is simply not true that everybody is doing it, with the inference that in some way fashion makes drug-taking an acceptable habit. Conversely, the fact of prohibition can also, regrettably, convey a spurious glamour and sense of excitement. The less harmful controlled drugs, cannabis especially, are widely available and relatively cheap. Group pressures to conform to what may be presented as a sophisticated way of life, embracing new and intensely pleasurable experiences, are strong. Fallacious analogies are made with smoking tobacco and drinking alcohol. Proselytizers, of whatever age, keen to prove that they are still in the swim, do incalculable harm by promoting the use of drugs. Younger addicts may switch from drugs to alcohol, or a combination of the two, thereby getting the worst of both worlds. For some, especially delinquent groups in the inner cities, drug-taking can be an expression of protest against established authority and a rejection of the outside world.

For many youngsters, first experiences fortunately do not live up to expectations. After experimentation, they may feel sick or ashamed or fearful of the consequences. Physical hangovers, worries about becoming caught up with pushers or hardened drug-takers, an awareness of friends who have been involved with the police or appeared in court, and reports of tragic incidents are all potent countervailing influences to the imagined attractions of taking drugs. Most authorities are in agreement that, of the very large numbers of young people who experiment, a majority soon decide that drugs are not for them and suffer no permanent harm.[36]

The idea of a cohesive and identifiable drugs culture in the sense in which it may have existed in the late 1960s no longer applies. Drug-taking has become much less of a deviant or fringe activity, now extending to people from all sections of the community and many geographical areas. This diffusion blurs the picture and invalidates some of the more extreme stereotypes. Yet the portents remain ominous. For some, the transient satisfactions of mood-changing drugs—the altered perceptions, the hallucinations, or the pleasurable feelings of relaxation or

[36] See *Drug Misuse—A Basic Briefing*, Leaflet DM3, issued by the Department of Health and Social Security and the Welsh Office. This leaflet was first published by the Institute for the Study of Drug Dependence in 1982, and was republished by the DHSS with the permission of the ISDD in March 1985. A revised edition followed in July 1985.

well-being—are a worth-while pursuit. They offer the promise of an easily obtainable alternative to a stress-inducing actuality, or a relief from boredom, loneliness, or inhibitions. Escapism of this sort may be a passing phase, co-terminous with another life unrelated to drugs; but equally, it can lead to withdrawal into a private world, with loss of self-esteem, fewer and less constant friends, and reduced motivation or expectations for the future. Having contributed to these shortcomings, drugs may seem to fill the gaps, but very often at the cost of even larger doses of increasingly powerful drugs, injected intravenously or inhaled to speed up their effect. The trend towards inhalation is inauspicious, avoiding the necessity of self-injection which many people find inhibiting. Whereas it may seem less harmful than injection, inhalation is capable of delivering at least as great an amount of the drug with no less hazardous effects.

Drug-taking by young people is a great leveller. Empty lives and unhappiness, or high-spirited experimentation, know no boundaries of social class or economic prosperity. Gifted children from good homes and with good prospects are attracted as well as deprived youngsters. The peer groups are different, but the attractions and the dangers are the same. Behind it lies the yearning for personal identity. Being someone, as a psychiatrist specializing in addiction behaviour has remarked, is very important for the young person who does not feel he knows who he is, what he is worth, or where he is going.[37]

The dangerous drugs which are controlled by law (of which heroin and its derivatives, cocaine, LSD, amphetamines, and cannabis are the most common) vary considerably in the extent of their misuse and the severity of their ill-effects, including the potential to produce psychological or physical dependence. LSD, for example, produces no physical dependence and may even work in ways that discourage frequent use. But cocaine, once mistakenly regarded as a fairly safe recreational drug, is now known to be highly compulsive and to induce strong psychological dependence.[38] Heroin stands out as inducing both

[37] Professor Griffith Edwards in *The Times*, 15 July 1986.
[38] The addictive quality of cocaine, at one time questioned, has been definitively established by some research carried out in the Bahamas by a medical team from the Department of Epidemiology and Public Health at Yale. See 'Epidemic Free-base Cocaine Abuse', the *Lancet*, 1 Mar. 1986, pp. 459–62. 'Crack', a more recent development, appears to be no more than cocaine hydrochloride.

physical and psychological dependence. It is of course true that tobacco is one of the most addictive of all substances, and that continuous long-term abuse of alcohol can lead to addiction: heroin, by contrast, is capable of inducing dependence in a few days or weeks, although typically it will build up over a longer period. At one time it was not usual to find young people starting off by experimenting with heroin: the drug was expensive, and its lethal potential was known and feared. But with the fall in its price from the beginning of the decade, and the switch from injection to inhalation of its vapour, heroin has become more attractive to some young people who have had little or no previous experience of illegal drugs.

The progression from experiment to intermittent abuse, and from regular abuse to addiction and dependence, although by no means automatic, is depressingly familiar to anyone who sees a large number of individual cases. Once addicted, it is very hard indeed to break the habit; and, apart from hospitals and a few residential centres where intensive therapy can help those with a strong desire to succeed, most of the medical treatment of addiction takes the form of either prescribing less harmful, although still habit-forming drugs, or conducting short-term detoxification backed up by counselling. Psychiatrists are usually responsible for in-patient and out-patient hospital treatment, since, even where addicts succeed in withdrawing from drugs, deeply engrained problems of psychological dependence may remain. A craving for the drug, a feeling of nostalgia for the drug-taking group, and a lack of anywhere else to go seem to be the main factors causing a return to drug-taking—as so often happens.

These generalities, so easy to state, conceal the particularity of countless incidents of degradation and collapse. Each is unique to the individual, but every one has repercussions on others. One example is that of a young woman, now aged twenty-eight and serving her second prison sentence, who had first been taken into care when she was thirteen. Not much was known about her family background, but she had been regularly truanting from school and was assessed as displaying behavioural problems. While in care she met a boy and twice became pregnant by him; she miscarried on the first occasion, and on the second she gave birth to a child who was brain-damaged and died at three months. The young couple continued to co-habit and in time were successful in obtaining a local authority flat. Three more children followed, two of whom were born after their mother had received a sentence of three months' imprisonment following her conviction on several charges of theft and deception. The offences were to get money, and consisted of tricking elderly pensioners to let her into their homes by saying that she was from the social services; once inside, she would steal their pension books and use them to obtain payments at post offices. This was her third appearance in the Magistrates' court, as she had previously been fined and conditionally discharged for offences of burglary and stealing from her employer.

There is no record of whether the court was aware at this stage that the young woman had been taking drugs since the age of sixteen. She had begun by smoking cannabis and taking hallucinogens. By nineteen she had experimented with barbiturates and heroin, and soon afterwards with amphetamines. In her early twenties, as the mother of three young children, she was injecting opiates regularly and was dependent on drugs. The young man whom she had first met when they were both in care had also become addicted to heroin and had accumulated convictions for drug-related offences. Before long he was to die, two days after his release from a prison sentence, from a drug overdose.

On her next appearances in court, which occurred on an approximately annual basis, the plight of the addicted mother (AM) was only too evident. Several of the offences were directly connected with drugs, including the cultivation of cannabis and six offences of forging prescription forms and attempting to obtain heroin substitutes from chemists. There were also more offences of theft. By now the probation service and social services department were mobilized; the children were removed into care and later sent to live with their paternal grandparents. In these circumstances the courts properly showed leniency, requesting and considering social enquiry reports, and on three separate occasions suspending sentences or making probation orders. Sometimes the supervision orders were linked to arrangements for medical treatment at an addiction unit, but these provisions came to nothing as the AM failed to keep in touch with her supervising officer or to attend the addiction unit.

Finally, she was charged with further offences of stealing pension books, in one incident roughly handling an elderly pensioner, although to her credit when he fell down she went back and helped him to get up. Once again, there were additional charges of theft and one of forgery, resulting in the AM being convicted on eight counts, with fourteen other offences taken into consideration. In breach both of her suspended sentence and probation order, the court decided it had no alternative but to impose a prison sentence.

This time the probation service did not ask for a lighter penalty. The report for the court stated that probation was unworkable since the AM's motivation was so low that she had failed to keep even one appointment. She had lost the tenancy of her flat after the death of her partner and the removal of her children, and was of no fixed abode. The probation officer's report concluded, exceptionally, that a prison sentence represented the only remaining hope: it was just possible that she might thereby be prevented from embracing death. The Crown Court to which the case was committed for sentence imposed two years' imprisonment on the substantive charges to run concurrently, with an additional fifteen months for the breach of the suspended sentence—a total of three years and three months imprisonment in all. The probation order was discharged. Leave to appeal against the sentence was granted, and the case was then heard in the Criminal Division of the Court of Appeal by the Lord Chief Justice and a Queen's Bench judge. Counsel for the appellant accepted the two-year sentences, and the sole ground of complaint before the court was that the suspended sentence ought not to have been brought into operation consecutively but should run concurrently with the other sentences of two years. The Court of Appeal did not accept this argument, nor that the total period of imprisonment that resulted was excessive. In the grounds of appeal, strong criticism was made of the probation officer's report—which it was noted, however, had not been challenged in the court below. In the judgment of the Court of Appeal, the boot was on the other foot. The report, they said, gave a convincing account of a persistent offender who had been leniently treated in the past but who had shown herself completely unable to co-operate with any agency or person who had tried to help her. Accordingly, the appeal was dismissed.

This unhappy sequence of events show how it is that a woman whose downfall has been drugs can come to be serving a prison sentence that to a casual observer might appear harsh and undeserved. In this case the fact of the AM's addiction had not only caused a complete breakdown in her personal and family relationships, but had left behind a trail of fear and loss among her elderly victims. Moreover, her own life was at risk. Nor is the dismal story yet ended. Medical opinion in her case was prepared to support release on parole only if her re-entry into the community was preceded by a period of up to eighteen months' residence in a therapeutic setting. The AM was not willing to agree; while she consented to receive out-patient treatment at a regional alcoholism and drug dependence unit located at a hospital close to her parents' home, she saw compulsory residential therapy as a continuation of her sentence. In our experience, said the doctors at the hospital, it was expecting too much of out-patient treatment: the AM had a great deal of learning to do which was unlikely to take place if she returned directly into the same environment from which she had come. There is force, too, in the comment made by one consultant that the decision whether to go to a residential centre was as much a matter for objective assessment as whether a young woman who had never enjoyed a normal drug-free adult life wanted to go or not. Thus free will clashed with professional expertise, not for the first or last time, and the former will probably prevail in the end. The probation service and social work agencies are indefatigably working on release plans, exploring possibilities with hostels and other alternatives. We must hope that this time their efforts will meet with more success. But the outlook is gloomy.

V

While it is possible to debate endlessly whether drug-taking should be regarded as an illness or a crime, the law makes the unauthorized possession, as well as the supply, of a controlled drug a criminal offence. The possession of small quantities of drugs for personal use in private is not invariably followed up by the police (if they have any way of knowing about it), since their limited resources are concentrated largely on investigating the sources of supply and distribution. At local level the police may refer misusers to other agencies for advice and treatment, but the fact remains that police enforcement results in numerous prosecutions for possession, often jointly with other charges, and that most of the 26,000 or so persons found guilty of or cautioned for drug offences in 1985 were takers rather than suppliers. Alarming as it is, the rapid increase in convictions over the last ten years in the trafficking, possession, and illegal importation of controlled drugs, which between them more than doubled in total between 1975 and 1985,[39] is only part of a wider criminality.

Drugs are costly, especially if taken regularly, and especially to those who may have lost their livelihood as a result of their addiction. Consequently, it is the need to finance the drugs habit, rather than the actual possession of drugs, which brings so many users into conflict with the law. An established addict's way of life comes to revolve around how to get sufficient supplies to satisfy his craving. Addicts are at the mercy of unscrupulous pushers, and to raise the money they frequently become suppliers themselves, introducing new members into the circle and so widening the distribution networks and spreading the evil. Women turn to prostitution, men to robbery, burglary, and theft from chemists' shops and pharmacies. Both sexes steal goods from shops which can be sold to raise the money to pay for what they so mistakenly see as their life support. Regional patterns vary, but the ramifications of drugs-related offending extend far and wide, and the cost to the nation is vast.

Although it is relatively easy to describe the problem, it is hard to say with any confidence what should be done about it. Whereas it is true that legal controls and prohibition patently cannot prevent the misuse of dangerous drugs, and in fact provoke part of the consequential offending, nevertheless, decriminalization is no answer. To aim for a free market in drugs, or in the less harmful drugs currently subject to legal controls, as

[39] About 25,000 persons were found guilty of, or cautioned for, offences involving controlled drugs in the United Kingdom in 1984. This was 1,700 (7%) more than in 1983 and 13,200 more than in 1975. *Tackling Drug Misuse: A Summary of the Government's Strategy* (2nd edn), HMSO, London, 1986, p. 5. In 1985 the total had reached nearly 26,000.

in the supply of alcoholic drinks, might have the effect of reducing the amount of associated offending, and limiting the opportunities for organized crime; but it is highly unlikely that it would take the money out of drugs or reduce their ill-effects. Even when methadone is lawfully prescribed as a substitute to heroin addicts, trading in it often takes place within hours of leaving the clinic, with money changing hands. Moreover, it is reported that new addicts in the catchment area of a treatment clinic frequently are found to have started on methadone. Even if it wished otherwise, no parliament or representative assembly in a democratic state, any more than an autocratic regime in a non-democratic one, would be permitted by public opinion to stand by and watch large numbers of its citizens destroy themselves. On purely medical grounds, the supply of class A controlled drugs including heroin, cocaine, and LSD, would in any event be strictly limited to prescription by medical practitioners, while it is most improbable that cannabis would get approval as a new medical drug today, much less for uncontrolled public sale. Thus, quite apart from the moral and political implications, any legalization of cannabis or other harmful drug would necessitate a dismantling of the regulatory framework for the manufacture and sale of medicines and drugs for pharmaceutical purposes.

The only practical alternatives are stricter enforcement of the law, international action to reduce supply, more treatment facilities, and educational and social policies aimed at prevention.[40] These are overlapping rather than exclusive objectives, although they do not hold out equivalent grounds for hope. In the United States, where drug abuse has become a mass phenomenon involving large numbers of adolescents and younger children, urgent attention has been directed towards identifying the most promising strategies to counter it. As part of its continuing criminal justice programme the Rand Corporation carried out an eighteen-month study reviewing the scale of adolescent drug use and the effectiveness of law enforcement,

---

[40] Reports were published by the Advisory Council on the Misuse of Drugs on *Treatment and Rehabilitation* in 1982 (DHSS) and on *Prevention* in 1984 (Home Office). Both were published by HMSO, London. The Council was established by the Misuse of Drugs Act 1971 with the duty to keep under review the problems of drug misuse and to advise Ministers on ways of dealing with them.

treatment, and prevention policies. The findings indicate that prevention is more likely than either law enforcement or treatment to reduce drug use among adolescents.[41] Despite legal controls, the researchers reported that drugs, especially marijuana (the Spanish name for herbal cannabis), are readily available to adolescents. National data estimated that 29% of high school seniors were smoking marijuana at least once a month and 6% were using it daily; 5% were using cocaine. The Rand survey concluded flatly that no feasible increases in law enforcement efforts would make any appreciable difference in drug availability or use. In order to cut the supply of all drugs enough to raise prices by 15%, investigative resources would have to be tripled. That represented an increase of $800 million, roughly the current total budget of the FBI. Even if expenditure on this scale were to be forthcoming, because marijuana is relatively cheap the investment would still not make it so prohibitively expensive for young people as to discourage its use.

The penetration by drugs into British society is not as deep— so far, at any rate—as in America. But our experience of more determined law enforcement has hardly been any more encouraging. Since July 1984, when the Government established an interdepartmental ministerial group on the misuse of drugs,[42] enforcement measures have received an enhanced priority. Police and Customs resources have been strengthened, fortified by a new National Drugs Intelligence Unit set up under a senior Scotland Yard officer in November 1985 to co-ordinate police and Customs intelligence in their efforts to combat the spread of illegal drugs. As a result of better intelligence and heightened vigilance at the ports, airports, container depots, and parcel post offices, increasingly large quantities of drugs have been seized. For example, a provisional total of 348 kilos of heroin was seized by the Customs in 1985, compared with 40 kilos in 1979. Yet all the indications point towards the continuation of drug smuggling, either accompanied or unaccompanied (through the post or in freight consignments), on a massive scale. No one can

[41] J. M. Polich, P. L. Ellickson, P. Reuter, and J. P. Kahan, *Strategies for Controlling Adolescent Drug Use*, Rand Corporation Report R-3076-CHF, 1984. *Criminal Justice Research at Rand* (October 1985) contains a summary of the research and lists the relevant publications at pp. 25–6.
[42] *Tackling Drug Misuse*, p. 7.

tell whether or not the quantities evading controls and coming into the country illegally are increasing at a faster rate than the amounts seized. However, in a trade about which so much is unknown or uncertain, the one unassailable index is that of the market-place: the selling price to the consumer, the so-called street price. The fact that the street prices of illicit heroin and cannabis resin have been relatively stable over the last five years, or have periodically come down (possibly reflecting more efficient distribution methods as organized crime has moved in), is a sure sign that supply is keeping up with demand. The existence of an expanding and price-stable market at a time of enhanced legal controls does little to inspire confidence in the efficacy of this response to the problem on its own.

If internal domestic controls, on the distribution and marketing of drugs as well as at the points of entry, are of only limited effect, the focus shifts to international co-operation to reduce supplies from abroad. This too is less unambiguous than it sounds. Whereas neighbouring countries facing similar drug problems tend to look on them in much the same way, being willing to share experience and co-ordinate enforcement measures (an example is the Pompidou Group in Western Europe), the producing countries are in a different situation. In some cases, notably Bolivia, a significant part of the national economy is dependent on the export of plants or substances from which cocaine is derived. Bolivia is now thought to produce about 40% of the world's cocaine, the value of the illegal crops being two to three times greater than the combined export income earned by all other products.[43] Where economic inducements and diplomatic pressures do not provide results, more drastic forms of intervention may follow. In Operation Blast Furnace in 1986, the Bolivian Government allowed American troops and agents from the US Drug Enforcement Administration to help the local drug enforcement unit to seek out and destroy cocaine-processing factories.[44] The British Government also concluded that international action to persuade peasant farmers to grow substitute crops was unlikely to be effective and, following a ministerial visit in September 1986, offered £600,000 to assist Bolivian police authorities with

---

[43] *Guardian*, 23 Sept. 1986.
[44] *The Times*, 8 Aug. 1986.

equipment and training. This contribution will be channelled through the United Nations Fund for Drug Abuse Control (UNFDAC), which has a law enforcement programme in Bolivia. Additional aid was offered to Peru and Ecuador.[45]

In the East, the cultivation of opium poppy crops may be integral to the way of life of large numbers of people subsisting on the growing of opium in poor soil for low rewards. They are surely among the least culpable of all the actors in the international drug trade. The social and cultural traditions in the producing countries will often be dissimilar from those in the more prosperous societies of the West. All this means that collective international action is hard to achieve. Programmes for crop eradication (the destruction of crops that have been grown illegally) do not, on their own, prevent the farmer from replanting at the earliest opportunity. The alternative is crop substitution (the growing of new crops in place of illegal crops), but that brings in its train far-reaching and costly implications. New crops may provide a livelihood, but they may also necessitate underpinning by a complete replacement infrastructure of rural development. Such economic intervention is not always welcome, as the Americans have found to their cost in South-east Asia and Latin America. Although there has been some useful bilateral co-operation between Britain and Pakistan, leading to such creditable gains as a reduction in the amount of opium produced in Pakistan from 800 tons in 1979 to about 45 tons in 1984,[46] the scale of the problem is universal and calls for a wider response. Finally, some provision has to be made for the fact that the opium poppy is the only raw material for the supplies of heroin and morphine that are required for authorized medicinal purposes.

Since 1961, United Nations conventions have sought to regulate the world-wide trade in drugs with the aim of ensuring that sufficient drugs are available only to meet legitimate needs. UN and other international agencies have co-ordinated projects to reduce the supply of drugs and to combat the growing

---

[45] Assistance from the United Kingdom amounting to approximately £1.5 million for drug-related activities in South America and the Caribbean was announced in October 1985, to be spread over three years. A large part of this aid is channelled through UNFDAC.

[46] *Tackling Drug Misuse*, p. 9.

criminality associated with the trade in illicit drugs, which sometimes reaches a pitch that threatens the stability of social and political systems. A fresh impetus was given towards collective action when the Secretary-General of the United Nations announced that an international conference on Drug Abuse and Illicit Trafficking will be held by the UN in Vienna in 1987. In the interim, the subject is one that is regularly on the agenda when heads of government and foreign ministers meet their counterparts, and is likely to remain so well after 1987.

Diplomatic initiatives supplement national policies and call for wholehearted support. In the long term, they may have an impact on the sources of supply (as has occurred in Pakistan) and on influencing attitudes in the producing countries. But it will be a long haul. One of the difficulties is that the hardship facing growers in developing countries, and the coming together of experts at international conferences, seem so remote from the immediacy of drugs abuse and the misery that accompanies it in the mainly urban areas in which it has proliferated.

VI

What expectations can be held out by treatment and rehabilitation of drug-takers? While there are glimmers of hope here and there, the prospects are not bright. First of all, since there are no proven methods of curing addicts or ensuring abstinence, it is hardly surprising that there should be so little agreement between practitioners about the best methods of dealing with drug abuse, or the exact nature of the problem to be treated. In particular, medical opinion is divided as to the wisdom of prescribing synthetic substitutes. Most adolescent drug abusers are not physically dependent when they come to notice; their recourse to drugs may be relatively recent, the result of peer group pressures, problems at school, or unhappiness at home. Most do not seek treatment, or even regard drug-taking as a problem for which they need help. But since few facilities are designed specifically for the treatment of adolescent drug-users, there is always a risk that those adolescents who seek them will gravitate towards units designed primarily for heroin addicts. Drug-users who are older and further down the road towards addiction and dependence are typically prepared to seek help only when they have become ill, or have appeared in court, or have been deprived of supplies. Even then, the treatment, if voluntary, may be accepted only until the immediate crisis has passed. It is a sad reflection that, unlike virtually all other types of patient, many addicts do not really want to be cured.

Clinics and rehabilitation centres may enjoy a measure of success with some addicts, and they deserve to be supported and extended; but, as with ex-addicts who are discharged from hospital or prison, the test comes on return to the temptations and pressures of life outside. It is a test that many fail: sometimes dramatically by overdose, sometimes more gradually, slipping back into the drug-taking habit by way of old friends and familiar places. There are dedicated people working heroically to help misusers give up drugs. When they succeed all can rejoice. Their work should be neither belittled nor ignored. There is certainly room for state intervention on the lines of the Central Funding Initiative of the Secretary of State for Social Services, which has made available over £17 million for pump-priming local projects providing services for drug-misusers. By August 1986, some 180 local projects in England had been aided under this scheme, ranging from clinics, drug screening equipment, and nurse training courses to counselling, telephone help-lines, and other services to assist drug-misusers and their families, some of them provided by voluntary agencies.

The present limitations to curative treatment have to be recognized. In the United States more than 3,000 separate treatment facilities for drug abuse in 1982 were supported by federal, state, and local governments, which provided 63% of the total cost of more than $500 million.[47] This sum represented a significant portion of national expenditure on all forms of drug control. After a systematic review of the relevant research and evaluations of effectiveness, the Rand report reached the carefully worded conclusion that the weight of evidence suggested that certain forms of treatment—namely, methadone maintenance, drug-free therapy, and residential therapeutic

---

[47] By 1986/7 it was reported that the total US budget for the overall anti-drug campaign will reach $3,200 million. Thirty-three federal agencies are taking part. *The Times*, 17 Sept. 1986.

communities—work better than no treatment at all.[48] It does no service, least of all to those who have a responsibility to decide on the thrust and direction of public policies, to sweep aside the realities of treating addiction.

VII

This leaves prevention—the most nebulous, but potentially the most promising, of the strategies reviewed in the Rand survey. Prevention programmes developed from some earlier and apparently effective anti-smoking programmes, based on a social influence model, have been tested in a sample of schools in California and Oregon since 1984/5.[49] The experiment is on a large scale and will continue for at least two years, following students through the seventh and eighth grades (ages thirteen to fifteen). The aim is to demonstrate to adolescents that there are socially acceptable ways of resisting the pressures towards drug-taking and smoking cigarettes and marijuana. Resistance skills— how to say no gracefully—are taught and counter-arguments developed. Students are helped to understand the physical and emotional costs of using drugs and the benefits of non-use. The American experience of drugs does not match our own exactly: the scale is greater, the educational system is different, and young people may react in ways that would not necessarily be reproduced in this country. But there will be other measures which are equally fruitful in influencing the attitudes, and subsequently, one must hope, the behaviour, of a generation that has so much at stake. Every avenue must be explored, and every approach exploited.

The complexities that lie behind drug abuse by young people mean that it is unlikely that any single measure, existing or new, will provide a breakthrough on its own. A comprehensive prevention policy is needed which not only applies to the schools but also provides for the dissemination of information about drugs to wider audiences. Since the establishment of the inter-departmental group of Ministers and officials, to which a

[48] Polich *et al.*, *Strategies for Controlling Adolescent Drug Use*, p. xii.
[49] P. L. Ellickson, *Project ALERT: A Smoking and Drug Prevention Experiment*, first-year progress report, Rand Corporation N-2184-CHF, 1984.

reference has already been made, the efforts of the Health and Education Departments, in Scotland and Northern Ireland as well as in England and Wales, have been intensified and brought more closely together under Home Office chairmanship. Educational projects have been devised and implemented, taking account of health education programmes related to alcohol and tobacco, and centred on the more specific aim of discouraging those who are not taking drugs from doing so. A campaign of national publicity directed primarily at young people likely to be at risk was launched in 1985/6, featuring television advertisements, posters, and advertising in the youth press.

Differences of opinion, sometimes strongly expressed, have revolved around the pros and cons of blanket advertising to undifferentiated audiences, the cost effectiveness, and the content of messages conveyed. Some critics, particularly those working in localities such as Merseyside, where there is a high incidence of drug misuse, have argued that, since the pattern is uneven across the country, publicity should be geared more to local needs. Others have questioned both the setting and the message. The theme is that taking drugs, especially heroin, is debilitating to personal health, endangers life, and drives away friends. Undeniable as this is, nevertheless, it has been alleged that the thin, drawn figures portrayed in the advertising have themselves come to exercise a certain fascination, with reactions reported from schools noting that some girls had put posters depicting the young male addicts on their bedroom walls because they found them attractive.[50] Clinical experience reinforces this occurrence, since, paradoxically, what repels the majority may (perhaps for that very reason) attract a minority. Although the numbers may be small, and so likely to be swamped in statistical evaluations, it is in this group that some of those most at risk are to be found.

Overall, however, the results of an independent evaluation commissioned by the Government indicate that the campaign had a measurable impact in increasing levels of awareness among young people about the dangers of heroin and in reducing the glamour associated with the drug. The most significant finding was that the percentage of teenagers sampled who said that they

---

[50]　*Sunday Times*, 13 July 1986.

would unequivocally reject an offer of heroin rose from 83%
before the campaign was launched to 94% after it had been
running a year. This is an encouraging statistic, despite the reality
that behavioural change does not automatically follow attitudinal
change; what people do does not always conform with their
stated attitudes. After reviewing the effectiveness of the
campaign, which also included leaflets and other material aimed
at parents and videos for use in school, the Whitehall departments decided to spend an additional £2 million in 1986/7
(equivalent to the cost in 1985/6) to develop it further. In the
second phase, addicts are depicted not as solitary and alienated,
but in social settings with groups of friends. Once again, the
selected target audience is young teenagers who are not
themselves drug-takers but are at risk of becoming attracted by
drugs.

A singleminded and non-sensational concentration on this
message—the damage to health and the ultimately anti-social
consequences of drug-taking—should remain at the heart of
prevention strategy. As time goes on, the proven ability of television, so often deplored, to induce behaviour that conforms to
what is presented as the norm can be expected to take hold,
particularly if similar attitudes and values are reflected in the
story-lines of fictionalized broadcast programmes. The aim must
be the creation of an anti-drug climate of opinion in society as a
whole, and there are some signs that this is emerging. But we
have to accept that the corollary may be a residue of high-risk
young people who reject the prevailing culture.

At the same time as national publicity campaigns, a structure
to enable local initiatives to be taken in response to localized
problems is taking shape. Funds have been earmarked to enable
each of the ninety-six English local education authorities to
appoint or second a full-time member of staff from 1 April 1986
to stimulate and co-ordinate action within the education service,
and in collaboration with other agencies, to combat drug
misuse.[51] On the health side, nearly all health authorities in
England have established drug liaison committees at district
level. Outside the co-ordinating machinery of the statutory
services, constructive work is done in many localities by self-

---

[51] *Tackling Drug Misuse*, p. 18.

help groups for drug-misusers and their families, backed up by prevention initiatives taken by the media, private industry, and organizations such as the YMCA and the Lions International.[52]

VIII

Unspectacular as much of this may be, nevertheless preventive measures offer the most fitting public response to the menace posed by drugs. Prevention is the most fitting because, alone of the responses open to the state, it gets close to the fundamental issue: why do people resort to drugs, and what can be done to keep them away from the perils of addiction? The law cannot resolve this issue. Despite strident demands for clamp-downs and tough action, hot-under-the-collar rhetoric is meaningless unless reflected by higher levels of enforcement. As we have seen, police and Customs activities have been strengthened, larger quantities of drugs have been seized, and more offenders have been prosecuted. Penalties have been increased, and life imprisonment is now the maximum sentence for trafficking in class A drugs. Steps have been taken to prevent those who traffick in drugs from profiting financially from their illegal activities; parole has been restricted; and the Court of Appeal has promulgated guidelines for more severe sentencing of those convicted of the most serious drugs offences. Through measures such as these, Parliament and the courts have demonstrated their condemnation of the drugs trade and those who are engaged in it. In so far as it is a public function of legislation and punishment to denounce or condemn, so signifying the gravity attached by the state to certain forms of criminal behaviour, this is a particularly striking example. And yet, the result is that, as more and more people follow one another into the courts, a proportion of them proceeding thence to add still further to the overflowing population of the prisons, the scale of the drugs problem has grown rather than diminished. This is not to argue that enforcement and penal sanctions have no place: clearly they do and they must. It is simply that they are incapable of providing a solution on their own.

[52] *Tackling Drug Misuse*, p. 19.

Treatment can ameliorate the problem and mitigate the damage that addicts are doing to themselves. Control of medical drugs, Customs action against smuggling, and international co-operation to reduce supplies are all practical forms of prevention if they are successful in stopping illicit drugs from entering the country or reaching the streets. The ready availability of such drugs and the level of prices they command, however, suggest that this form of physical prevention still has a long way to go before it begins to bite. Indeed, as external controls become more effective and prices rise, one consequence to be anticipated is the prospect of increased levels of criminal offending by addicts to obtain more money to pay the higher prices.

Nothing in this chapter is intended to minimize the consequences or seriousness of addiction to other toxic substances, notably alcohol, which are not necessarily harmful in themselves but are capable of causing infinite harm and distress if taken to excess or otherwise misused. The relationship between alcohol and crime is far-reaching and has been the subject of much study.[53] Yet, although the agent is different, as are the social dimensions and public perception of the problem, current policies towards the treatment and prevention of alcoholism and drugs are remarkably similar. This is as it should be, since the issues and the policy responses have more to do with people than with the substances that are abused. So we come back to the mind of the individual, particularly the young person who, alone or in groups, is attracted towards drugs, and how he can be influenced. References are sometimes made to a drugs epidemic and the analogy is an apt one. In Britain youthful drugs misuse first manifested itself on a national scale in the 1960s, with an upsurge in the late 1960s and early 1970s. A period of relative stability followed in the mid-1970s, but since 1979/80 all of the indicators have soared up again.

That is the way with epidemics: the outbreak, at first unnoticed; the slow start; the arguments over symptoms and causes; the failure of treatment; the gradual build-up; perhaps a temporary check; the false hopes preceding the sudden onrush;

[53] R. Hauge, *Alcohol and Crime*, Council of Europe, Strasbourg, 1984. See also a report of a conference held in London in June 1986: D. Acres, 'Alcohol–Drugs–Crime', the *Magistrate*, 42 (1986), 136–7. Dr Acres is Chairman of the Council of the Magistrates' Association.

the public panic at the apparently uncontrollable flood. But it is also in the nature of epidemics, whether contagious or non-contagious, that they eventually run their course. That mankind has survived as a species is evidence that even the most devastating of epidemics, which have rapidly spread to afflict large numbers of people simultaneously, at some stage and for reasons that may not be perfectly understood, stabilize before going into decline, with the incidence of the disease reverting to somewhere near its former level. In the end, the remarkable resilience of the human race prevails. Will the same be true of the epidemic of drug addiction, essentially a disease of the spirit rather than the body, which has swept across so much of the surface of the globe, especially in the most highly developed countries, in the closing years of the twentieth century?

7

Prison: The Last Resort

I

I find it hard to write about prisons and those who are imprisoned, harder than any other chapter in this book. At first one tends to think, in tune with much reformist opinion, that far too many offenders are sentenced to imprisonment, and for too long. The prison system itself, the way it is organized, and the sentencing process receive less scrutiny. A trenchant criticism, now well aired, is that Britain sends more people per head of the population to prison than almost any other European country. Unflattering comparisons are drawn and jibes made, even by a former Home Secretary of the stature of Roy Jenkins: 'At least we lead Europe in the ... size of our prison population, the Turks having given up that race in despair.'[1] Including remand prisoners, the prison population per 100,000 inhabitants in February 1986 was 94 for the United Kingdom and 92 for England and Wales. Compared with other countries in membership of the Council of Europe, this proportion was broadly comparable with West Germany but below Austria (109), with Turkey still heading the table at 130.

In practice, international comparisons are inexact and not particularly reliable as indicators of national policies or attitudes to imprisonment. None the less, the United Kingdom undeniably has a relatively high prison population calculated in terms of the numbers in prison as a proportion of the population as a whole, although it does not follow that is necessarily a matter calling for apology or excuse. As to the length of sentences, it does not

[1] From a speech winding up a debate on freedom, democracy, and better government at the SDP Conference, Harrogate, 15 September 1986. Few Home Secretaries, past or present, or leading spokesmen in other parties, have found party conferences a comfortable platform for the exposition of penal policy. The official statistics on prison populations in member states published by the Council of Europe do not bear out this claim.

appear that the average periods of detention served by those who receive custodial sentences are exceptionally long by European standards, whereas in the United States, if those in custody in local gaols are included as well as sentenced prisoners in state and federal institutions, the rate per 100,000 resident population is about three times higher than in the United Kingdom.[2]

Confining ourselves to England and Wales, the subject of this book, the postwar trends of imprisonment have been steadily upwards. In 1947 the average daily population in custody was 17,067, made up of sentenced prisoners (including young offenders), remand prisoners awaiting trial or sentence, and a small number of civil prisoners. By 1957 the figure had risen to 22,602, and ten years later, in 1967, it reached 35,009.[3] A detailed profile of the prison population between 1972 and 1985 is contained in Table 12. This gives the actual and percentage figures for each of three categories of reception (sentenced, remand, and non-criminal) and shows the principal offences that have attracted custodial sanctions. The totals demonstrate an escalation in the population as measured on 30 June in each of the years listed, from 38,582 in 1972 to 47,503 in 1985. There are considerable fluctuations within the course of each year; for example, during the first half of 1985 the population in custody surged from 41,400 at the end of the previous year to a peak of 48,200 at the beginning of August.[4] Thereafter it subsided, standing at 47,200 at the end of September 1985, some 1,000 less than two months before. The average daily population in custody throughout 1985 was 46,200.

There is much to reflect upon in Table 12. First, and perhaps surprisingly, we should note that the number of offenders serving sentences of imprisonment for non-violent property crimes has

[2] The detention rate per 100,000 resident population in the USA for 1983 was 277, made up of 179 sentenced prisoners in state and federal institutions and 98 sentenced and unsentenced gaol inmates. E. F. McGarrell and T. J. Flanagan (eds.) *Sourcebook of Criminal Justice Statistics—1984*, US Department of Justice, Bureau of Justice Statistics, Washington, DC, US Government Printing Office, 1985, pp. 638, 647. The total gaol and prison population in the USA increased from about 260,000 in 1950 to 630,000 in 1985.

[3] *Committee of Inquiry into the United Kingdom Prison Services, Report* (Cmnd. 7673), HMSO, London, 1979, p. 31 (the May Report).

[4] *Prison Statistics England and Wales 1985* (Cmnd. 9903), HMSO, London, 1986, p. 7.

TABLE 12. *Population profile of Prison Department establishments, by offence group and type of prisoner, 1972–85*

	1972		1976		1980		1984		1985	
	No.	%	No.	%	No.	%	No.	%	No.	%
Sentenced										
Violence against the person and sexual offences	6,342	16	7,733	19	8,582	20	8,151	18	8,839	19
Robbery	2,389	6	2,190	5	2,362	5	2,694	6	3,028	6
Burglary, theft, handling, fraud, forgery, arson, and criminal damage	21,917	57	22,682	54	21,797	50	19,460	44	19,764	41
Other and not recorded[a]	2,801	7	3,670	9	3,896	9	5,191	12	5,713	12
TOTAL SENTENCED[a]	33,449	87	36,275	87	36,637	83	35,496	80	37,344	79
Remand	4,653	12	4,934	12	6,708	15	8,626	19	9,939	21
Non-criminal	480	1	495	1	591	1	311	1	220	0
GRAND TOTAL	38,582		41,704		43,936		44,433		47,503	

[a] Including fine defaulters.

declined, both numerically and as a percentage of the annual totals. This trend, however, has been offset by the rise in the number of prisoners serving sentences for offences of violence against the person, including sexual offences, and robbery. The greatest increase is to be found in the remand prisoners, who are either awaiting trial or, having been tried and convicted, are awaiting sentence. Here the totals have more than doubled, from just over 4,600 in 1972 to close on 10,000 in 1985. Although the numbers are small, there has been a reduction in the total of non-criminal prisoners resulting from civil proceedings by more than a half over the same period.

In the last chapter the dramatic reduction in the size of the prison population in the first quarter of the twentieth century was cited, and this is something that is still brought up in current debate.[5] Why could it not be done again? Is it the determination, the originality of mind, the vision that is lacking? Or has the situation been transformed to an extent that prevents meaningful comparison? How has it come about, at a time of severe restraint on public expenditure, that spending on prisons, the least popular of all objects, has increased more rapidly than that on schools, hospitals, and the social services? Above all, would the successors of the public that acquiesced in Churchill's reforms reject any similar initiatives today?

The right place to start is with the rudimentary question: why send people to prison at all? If some offenders are to be sentenced to terms of imprisonment and others to non-custodial penalties, on what grounds is the distinction to be made? Throughout the discussion, we should remember that to deprive a man or woman of their liberty, of the right to move about or conduct their private lives as they wish, is the most extreme restriction that can be imposed in a free society. Only those who have experienced it can appreciate the full significance of being deprived of the most ordinary and taken-for-granted aspects of everyday living.[6] The fact that most prisoners adjust, and come to accept their lot, says more about human adaptability

[5] See A. Rutherford, 'Prisons: Follow Winston's Lead', *The Times*, 20 Aug. 1985.

[6] A well written recent account of prison as experienced by prisoners is James Campbell's *Gate Fever: Voices from a Prison*, Weidenfeld & Nicholson, London, 1986.

than about the extent of deprivation. It is unfortunately a truism that the regimes to which modern-day prisoners are subjected, despite the best efforts of officialdom to the contrary, are unavoidably harmful. Occasionally, as in the case of the young woman heroin addict recounted in the last chapter, a prison sentence may save a life, but such instances are exceptional and almost incidental. Imprisonment cannot properly, in principle or in practice, be used as a check on the consequences of self-abuse.

The overcrowding experienced in recent years makes worse everything that is worst about prison. The most senior administrators in the penal system admit publicly that serving a sentence in the overcrowded conditions which have become only too familiar is very different from serving a sentence in more relaxed and uncrowded conditions.[7] Privacy, exercise, mobility within the prison, visits, industrial training and education—the very things that make prison bearable to those inside and tolerable to those outside—all suffer. Overcrowding also induces tensions and fears of violence or riots, which breed repressive staff attitudes. None of this is reflected in the sentence awarded by the courts; indeed, as public opinion has hardened in the direction of tougher responses to crime, more offenders have been sent to prison and sentences have got longer. Table 13 shows clearly enough the increased use of imprisonment by the courts, a trend that is currently moving against the counter-cycle of a fall in the total number of offenders convicted and sentenced for indictable offences in the Crown Court and the Magistrates' courts. This may reflect the greater use of cautions by the police, which has been especially marked in the case of young people. The decline in the proportionate use of fines is also significant.

Table 14 is also salutary. It demonstrates that, although the sentenced population in the prisons has not markedly risen over the five years from the start of 1981 to the end of 1985, the sentence lengths for almost every category of offender, women as well as men, have increased.

[7] See evidence given by Sir Brian Cubbon, Permanent Under-Secretary of State at the Home Office, to the House of Commons Committee of Public Accounts, *Twenty-fifth Report, Prison Building Programme*, Session 1985–6 (HC, 248), HMSO, London, 1986, p. 7.

TABLE 13. *Convicted offenders sentenced for indictable offences, 1977–85*

	1977	1979	1981	1983	1985
Total no. of offenders convicted and sentenced (custodial and non-custodial) at all courts (thousands)	426.2	412.0	464.4	461.8	444.3
Sentenced to immediate imprisonment[a] (including partially suspended sentences) (thousands)	55.1	57.9	69.3	73.6	78.2
Percentage of total	12.9	14.1	14.9	15.9	17.6
Fines (£'000)	218.1	206.7	209.9	199.3	177.7
Percentage of total	51.2	50.1	45.2	43.1	40.0

[a] Including detention centres, youth custody, or Borstal training.

II

Imprisonment is expected to serve several purposes simultaneously. It is seen as a punishment for wrongs done, the most grievous sanction available to the courts. It functions as a reassurance and conduit for public feelings of anger and outrage; and it is a method of ensuring that persistent or high-risk offenders are prevented from committing further crimes against the public while confined in custody. These objectives do not always coincide, but they combine to condition our thinking about who should go to prison and how long they should stay there.

The undifferentiated mass of offenders found guilty of criminal offences can be reduced to a small number of definable groups. First, there are those whose crimes are regarded by the sentencing court as being so objectionable that they must be punished by the most severe penalty that can be imposed. The length of the sentence of imprisonment will be related to the gravity of the offence, with allowances made for mitigating or aggravating factors. But even a first offender, hitherto of good

TABLE 14. *Sentenced population of Prison Department establishments, by sex, type of offence, and sentence length, on 30 June 1981 and 1985*

	Up to 2 yrs[a]		2–4 yrs		Over 4 yrs		Life		Total	
	1981	1985	1981	1985	1981	1985	1981	1985	1981	1985
Males	23,708	22,003	6,403	7,193	3,812	5,000	1,626	1,991	35,549	36,187
Violent offences[b]	18,576	17,560	4,033	4,438	1,503	1,986	1,573	1,928	24,165	24,047
Other offences and not recorded	4,253	3,903	2,370	2,755	2,309	3,014	53	63	10,505	11,600
Fine defaulters	879	540	—	—	—	—	—	—	879	540
Females	913	861	114	159	44	77	49	60	1,120	1,157
Violent offences[b]	139	112	80	124	27	49	48	59	238	234
Other offences and not recorded	735	726	34	35	17	28	1	1	843	900
Fine defaulters	39	23	—	—	—	—	—	—	39	23
Males and females	24,621	22,864	6,517	7,352	3,856	5,077	1,675	2,051	36,669	37,344

[a] Including Borstal trainees.
[b] Violence against the person, sexual offences, and robbery.

character, who, for instance, pleads guilty to a serious offence of violence to the person will have to expect an immediate prison sentence. Next come those who were previously given a chance, but failed to take it. A substantial number of prisoners have demonstrated by their behaviour that non-custodial penalties, such as probation, fines, or community service orders, have been ineffective in discouraging the commission of further crimes. They too will have to anticipate a period in prison. This category overlaps to some extent those who represent a continuing risk of persistent offending (e.g. recidivist burglars), or of conduct which could lead to grave losses of life or property (e.g., compulsive arsonists). Recidivists make up a large group in prison and present particularly difficult problems when it comes to considering their suitability for early release.

Standing somewhat apart from these classic criminal offenders are those who have consistently or deliberately refused (for whatever reason) to pay fines or conform to other orders of the court. Although numerous as a category defined in terms of reception into custody (20,491 in 1985),[8] fine defaulters have only a marginal effect on the overall prison population since they seldom spend more than a few days in custody.[9] Thus, for example, on 30 June 1985 there were no more than 563 fine defaulters in custody.[10] An even smaller, but in some ways more intractable, group is made up of conscientious objectors or people who feel obliged towards civil disobedience as a method of demonstrating their point of view. Domestic disputes, leading to orders for maintenance payments or injunctions to keep away from former spouses, co-habitants, or children, can also result in custodial sanctions. As already noted, the number classified as civil or non-criminal prisoners has declined since 1980.

A majority of indictable cases are punished by the imposition of non-custodial penalties. Of the 444,300 persons shown in Table 13 as convicted at all courts in 1985, immediate imprisonment was found to be necessary in 17.6% of cases. The figures

[8] The 1985 total was about 1,300 fewer than in 1984 and 4,000 fewer than in the peak year of 1982. *Prison Statistics England and Wales 1985*, p. 89.

[9] The length of time served on average by fine defaulters received into Prison Department establishments in 1985 was about 11 days for males and 8 days for females. Ibid., p. 89.

[10] See Table 14.

were considerably higher at the Crown Court, where about one-third of those convicted were given a sentence of immediate custody, and a further 18% a partially or wholly suspended sentence of imprisonment. At the Magistrate's court the proportionate use of imprisonment was much lower.[11]

The availability of supervision and guidance provided by the probation service, especially for inadequate offenders who find it difficult to cope with the pressures of their lives, may influence the sentencing court. The standard of supervision provided under probation and community service orders will affect the court's confidence in such orders and its readiness to use them in cases where custody might otherwise result. If the offender is in need of a controlled and supportive environment, conditions of residence may be added to a probation order. For juveniles, a supervision order with a condition of intermediate treatment can take the place of a detention centre order.

Apart from these convicted offenders (and civil judgment defaulters), there is another category which swells the over-flowing prison population still further. These are the so-called remand prisoners, technically not offenders at all, but accused persons awaiting trial or sentence and with privileges such as wearing their own clothes and an entitlement to daily visits to mark their separate status. Since the Bail Act 1976, all accused but unconvicted persons have a right to bail, unless in the judgment of the court there is a likelihood that they will fail to attend the court as required, or that they may commit further offences, or interfere with the course of justice, if remaining at liberty. In reaching its decision, the court may take into account the nature and seriousness of the charge against the accused, as well as his character and previous record. The result is that, out of a total population in the prisons close on 47,500 at the end of June 1985, nearly 10,000 were untried prisoners who had been remanded in custody. This large category, about 21% of the whole, is partly the consequence of more receptions, but is also affected by the longer time that offenders are held on remand. Delays are caused not only by more detected offenders being brought before the criminal courts, but by the fact that more of them are being committed to the Crown Court for trial. Since

[11] *Criminal Statistics England and Wales 1985* (Cm. 10), HMSO, London, 1986, p. 135.

delays in the Crown Court are usually greater than in the Magistrates' courts, a backlog builds up which is soon felt in the prisons. Of the untried prisoners who were remanded in custody during trial at the Crown Court, a relatively small proportion, amounting to some 6% in 1985, were acquitted, with a further 11% receiving non-custodial sentences. But the problem about simply stating that far too many people are remanded in custody is finding a way of sorting out the sheep from the goats. I shall have more to say about the detention of remand prisoners later in this chapter.

If most of the sentenced offenders falling within the categories mentioned above are sent to prison for a period of time that is regarded by the court as equating with the 'just deserts' earned by their criminal behaviour, a greater conceptual difficulty arises with those who are imprisoned mainly because they represent a danger to the community. While the commonsense justification of protecting the public is plain enough, the fact remains that these offenders are treated in a way that is different from others, who may have committed identical or broadly similar offences. Here the pure doctrine of retributive justice, apparently so symmetrical and satisfying, fails to meet the circumstances presented to the sentencing court. Some flexibility is required in the sentencing process to allow for previous conduct and the likelihood of re-offending to be taken into account. Both in sentencing decisions, where one man may be sent to prison for an offence that has earned another a non-custodial penalty, and in decisions about early release, where one man may be granted parole which is denied to another serving a similar sentence, pragmatic judgments are made whenever an assessed proclivity to dangerousness exceeds the punitive sanction earned by the crime. The best way to approach issues of this sort is to distinguish carefully between two descriptions which are often used interchangeably: *incarceration*—a sanction measurable in relation to the offence; and *incapacitation*—keeping an offender out of harm's way while it is thought he is still a danger to others.

A theorist as closely identified with doctrines of retribution and deterrence as Ernest van den Haag has admitted that, 'Although persuasive, the argument for keeping an offender in prison because he is still dangerous leaves one uneasy.' He asks whether we can:

keep a man in prison solely because we think he will commit crimes again if released? If punishment is to be given only for what is deserved by what the convict has done in the past, how can he be punished for what is feared of him, for what he may do in the future? We cannot punish offenders just to protect society from anticipated danger. 'Punishment' refers only to what is deserved for a crime already committed. Nobody can be punished for being dangerous or inclined to violence. Character—a set of dispositions to act—cannot be punished; only acts can be.[12]

Although both can lead to custody, the distinction between punishment and the protection of the public is crucial. Van den Haag himself observes that this rigorous interpretation of justice is not the only public good, and that the preservation of society and the security and welfare of its members are legitimate political ends beyond justice. Psychotics are confined, and persons with infectious diseases quarantined, not because they have broken any law, but because their continued presence in the community at large constitutes an excessive hazard. On this reasoning, once a convicted offender who has been sentenced to imprisonment has served the minimum term required by law for his crime, he may either be released, or continue to be incapacitated if there is reason to believe that he will unlawfully harm others if released. In such cases the potential for future harm, and the likelihood of it occurring, outweighs the harm done to the individual by prolonging the restrictions on his freedom.

The practical difficulties lie in the assessment of risk and identifying symptoms of dangerousness, a condition that is questioned by some experts. 'Because we have all experienced dangerous emotions,' wrote the late Dr Peter Scott, 'there is a risk of projection and scapegoating. The label, which is easy to attach but difficult to remove, may contribute to its own continuance . . .'.[13] This does not mean, as is sometimes claimed, that dangerousness is an unscientific concept. In accepting that it is certainly imprecise, in that it does not convey the nature of the

[12] Ernest van den Haag is John M. Olin Professor of Jurisprudence and Public Policy at Fordham University, New York. His provocative, but influential, book *Punishing Criminals* was first published by Basic Books, New York, in 1975. See pp. 242–3.

[13] P. D. Scott, 'Assessing Dangerousness in Criminals', *British Journal of Psychiatry*, 131 (1977), p. 127.

danger, Professor Nigel Walker has denied that it is unscientific to attribute fairly specific sorts of propensity to individuals, such as personal violence or paedophilic molestation. Walker goes on:

It may well be unscientific to infer such a propensity from a single incident; but so long as there is evidence of repetition, and so long as it does not lead to an assumption that the propensity will be actualised in response to every temptation or provocation, it is not unsound. If it were, we would be debarred from acting on the assumption that people are liars, or truthful, bad or prompt payers of debts, responsible or irresponsible parents.[14]

That these considerations are more than simply academic is seen in the distinction between 'tariff' and 'risk' cases which is used by the Home Office, the higher judiciary, and the Parole Board in deciding on the release of life sentence prisoners on licence. When consulted as to the requirements of retribution and deterrence (i.e., the tariff), the judiciary, in the shape of the trial judge (if still available) and the Lord Chief Justice of the day, may advise that these criteria have been satisfied after a period of time has passed and that the case falls only to be considered from the standpoint of risk. A further discussion of these procedures will be found in the next chapter.

III

Although the available research indicates that there does not seem to be much difference between custodial and non-custodial sentences in terms of comparable offenders' subsequent conduct,[15] prison does not exist simply as a way of dealing with the problems of recidivism. There are the wider functions of denunciation and punishment, as well as the protection of the public. In modern society a prison system of some sort is inescapable, and its occupation is determined largely by the sentences passed by the independent judiciary. That there should be a divergence of interest between those who impose the sentences and those who have to provide the necessary prison

---

[14] 'Unscientific, Unwise, Unprofitable or Unjust?', *British Journal of Criminology*, 22 (1982), 276.
[15] R. G. Hood and R. Sparks, *Key Issues in Criminology*, Weidenfeld & Nicolson, London, 1970, p. 215.

places is a natural consequence of separating the powers of the state and as such is fundamentally healthy. Proper though this may be to the constitutionalist, it raises acute practical problems for the administrator who is condemned to the fruitless task of having to respond to a demand over which he has no control.

Independent as it is from political direction, the judiciary is subject to the public mood. Judges and magistrates live in the community and recognize that the courts form part of the way in which social relationships are regulated. Sentencing decisions are taken on the facts of each individual case, but with an awareness of public expectations of retribution and reassurance. In the first of the 1986 Reith Lectures, Lord McCluskey remarked on the judge's consciousness of the wider audience:

He knows that society is watching him. And he knows that he alone, at that moment, is charged by society with the power to mark, by the sentence he imposes, its sense of outrage, tempered by the justice of the case, by the wickedness and the frailties of the offender, and by the inability of the penal system to help either the offender or the victim in any meaningful way. He must disregard the fact that almost any sentence of imprisonment he imposes may be materially, but unpredictably, altered by administrative decisions which will take effect, years later, in circumstances he cannot foresee. He can pronounce sentence in seven words or he can give an elaborate explanation of why he is doing what he does. He can deliver a sermon on the human condition. He can deploy his extensive vocabulary of words of censure, pity, revulsion, warning and threat. But, one way or another, from some inscrutable depths of his own gut, he has to make a choice which no rule of law compels.[16]

Although the principle, enshrined in statute,[17] remains that a sentence of imprisonment should be imposed only in the last resort and where no other penalty will suffice, it has not

---

[16] Lord McCluskey, Reith Lectures; reprinted in the *Listener*, 6 Nov. 1986, p. 16.

[17] Powers of Criminal Courts Act 1973 S. 20 prescribes that no court shall pass a sentence of imprisonment on a person of or over 21 years of age on whom such a sentence has not previously been passed unless the court is of the opinion that no other method of dealing with him is appropriate. Under S. 1(4) of the Criminal Justice Act 1982, a court must not send a young adult offender to detention centre or youth custody unless it is of the opinion that no other method of dealing with him is appropriate because it appears to the court that he is unable or unwilling to respond to non-custodial penalties, or because a custodial sentence is necessary for the protection of the public, or because the offence was so serious that a non-custodial sentence cannot be justified.

prevented a greater recourse towards custodial sentences over the last decade. This trend between 1977 and 1985 was clearly shown in Table 13. For a ten-year comparison, whereas 50,000 people were convicted of an indictable offence in 1975 and sentenced to immediate custody, the number had risen to over 78,000 in 1985.[18] In 1975, 16% of the men over twenty-one who were convicted and sentenced for indictable offences received immediate prison sentences: by 1985 the figure had risen to 21%. Similarly, in 1975 only 3% of the women over twenty-one who were sentenced for indictable offences were sent to prison immediately, a statistic that had more than doubled by 1985.[19] While the Home Office takes care to stay at arm's length from sentencing decisions, it can and does offer guidance to the courts about the content and effect of the various sentences at their disposal. On the use of imprisonment the official handbook, entitled *The Sentence of the Court,* is explicit:

It would be wrong to impose a custodial sentence in a case where that severe a penalty was not warranted by the crime in question, merely in the hope of achieving a deterrent or reformative effect which experience suggests is unlikely to materialise. It is indeed important to bear in mind that custody is not only the most severe penalty available, and ... the one which consumes most resources, but it is also the one likely to have the most serious side-effects. These include not only harm to the offender's personal and financial prospects on release but also the particular likelihood of inflicting hardship on the offender's family as well, rather than merely exacting proper retribution from the actual criminal. All these considerations are summed up in the general principle, laid down by the Court of Appeal, that custody should only be imposed when it is truly necessary in the circumstances of the case, and that if it is necessary, the sentence should be as short as is consistent with the need for punishment.[20]

What could be clearer? The reference to resources brings up the financial cost of imprisonment to the taxpayer, as well as the social cost to the prisoner and his family. A revealing table published as an appendix to the handbook discloses the relative costs of alternative sentences.[21] Whereas an adult male offender

[18] *Criminal Statistics England and Wales 1985,* p. 115.
[19] Ibid., p. 149.
[20] *The Sentence of the Court: A Handbook for Courts on the Treatment of Offenders,* HMSO, London, 1986, pp. 7–8.
[21] Ibid., Appendix 3, p. 101. The basis of the calculation was the average cost per offender per week in 1982/3.

in a high-security prison cost the state £433 per week, the average weekly cost in other closed training prisons was £205, and in open prisons £158.[22] Against this, the average cost of a community service order was £10 per week, a probation order £11, and a supervision order £9. Despite the adverse factors of cost, harmful effects on offenders' families and reinforcing the criminality of convicted offenders, prison sentences continue to be passed, sometimes, it seems, almost as a matter of routine.

There is no evidence to suggest that longer periods of custody lead to reduced re-offending; nor do research findings indicate that the imposition of particular sentences, especially severe sentences, has any more than a very limited effect on crime levels.[23] Few people, sentencers or sentenced, are persuaded of the deterrent effect of imprisonment, and fewer still believe that a prisoner may be reformed as a result. In most cases custodial sentences are imposed for no better reason than they seem to the court to be unavoidable: the offence is a serious one, or the offender has been before the courts on several previous occasions when non-custodial penalties were tried and failed. Less directly, the idea of imprisonment as the embodiment of severity in punishment, is in fashion—although more in the abstract than when related to individual cases not involving violence.[24] Neither judges nor magistrates relish the prospect of being vilified as over-lenient or insufficiently zealous in their allotted role of enforcing the law. And so the numbers in prison climb on upwards, future projections are made and revised, and new prisons are rapidly filled to their planned capacity soon after they open.

[22] In 1985/6 the average weekly cost of custody was £470 for an adult male offender in a high-security prison, £212 in other closed training prisons, and £157 in open prisons. The cost of non-custodial sentences had also risen, to £12 per week for a community service order, £14 per week for a probation order, and £11 per week for a supervision order.

[23] *The Sentence of the Court*, p. 6.

[24] On the basis of a survey commissioned by the *Observer* in 1982, with a national sample of 988, the Prison Reform Trust claimed that popular opinion was less punitive and more sympathetic towards penal reform than had previously been believed: Stephen Shaw, *The People's Justice*, Prison Reform Trust, London, 1982. See also a NACRO briefing paper on *Public Opinion and Sentencing*, published on 1 December 1986, which reached the same conclusion.

IV

The attempt to keep up with the tide may be a near impossibility, but the state of the prisons precludes any question of its abandonment. Overcrowding has enhanced the already sordid conditions in what are too often decaying and obsolescent buildings. Lack of integral sanitation is only the most conspicuous of the many inadequacies that disfigure much of the prison estate, especially in the older local gaols in the city centres, where the weight of numbers is most felt. It is these conditions, fuelled by forecasts of ever-rising numbers of sentenced prisoners, that have led to the biggest building programme since the Victorian era. In the forty years between 1918 and 1958, helped by a static or falling crime rate between the wars, no new prisons were built. As the prison population continued to rise inexorably throughout the 1970s and early 1980s, the then Home Secretary, William Whitelaw, fortified with the recommendations of a Committee of Inquiry into the UK Prison Services,[25] convinced his Cabinet colleagues that a substantial prison building programme was imperative. This commitment was inherited by his successor, Leon Brittan, who came to the Home Office after the 1983 general election from the Treasury where he had been Chief Secretary, and who was able to obtain sufficient ministerial support to expand and accelerate the programme.

Its dimensions were ambitious: the construction of fourteen completely new prisons and major capital works of improvement at no less than 90 out of the then 120 Prison Department establishments, including twenty-three out of the twenty-five local prisons. It was carefully explained that the total number of additional places in the new prisons would not be a net gain, since some existing accommodation would be lost when it was taken out of use to be refurbished. The programme was none the less sufficiently extensive for the Government to announce a target of ending overcrowding by matching available places with the estimated prison population by the end of the decade. Despite criticisms of the way in which the building programme has been implemented, the subject of a report by a House of

---

[25] *Committee of Inquiry into the United Kingdom Prison Services, Report.*

Commons Select Committee in the 1985/6 Parliamentary Session,[26] the Home Office cannot be faulted for lacking a sense of urgency. Three newly built prisons were opened in 1985, with the addition of a former RAF camp at Lindholme which was acquired, converted, and brought into use in less than four months: it is now a permanent facility providing 750 places. Six other prisons are under construction and should be ready for use in 1990, with nine more in various stages of planning and design. Design resources for a further two prisons are committed for 1987/8, bringing the total number of new prisons, completed or projected, to twenty-one. A programme of this size is expected to provide about 10,600 new places.[27]

The cost of this huge undertaking—renovating numerous outdated structures and modernizing the living accommodation in most of the existing prisons as well as building completely new ones—is formidable. Expenditure on the construction of the new prisons was estimated in 1986 at well over £400 million, with the cost of major capital works of improvement in progress or planned at the existing establishments expected to reach over £700 million by the end of the century.[28] The drive to provide extra places should help to improve conditions and reduce over-crowding; although it now seems unlikely that, as had been hoped, the necessity for two or three prisoners in certain local prisons to share a cell designed for one will be eliminated by the end of the decade. Worst of all, the Home Office anticipates that the humiliating and degrading practice of 'slopping out' chamber pots that have been used during the long night hours when an inmate is locked up in the confines of a small cell, sometimes exercising his bodily functions in the presence of one or two other prisoners, will still apply to between 10,000 and 20,000 prisoners in 1991.[29]

Prison standards, overcrowding, and public expenditure are closely linked and dependent on one another. It is very hard to

[26] House of Commons Committee of Public Accounts, *Twenty-Fifth Report.*
[27] *Criminal Justice—A Working Paper* (rev. edn), Home Office, London, 1986, p. 29.
[28] Ibid., p. 30.
[29] Evidence given by Sir B. Cubbon to the House of Commons Committee of Public Accounts *Twenty-Fifth Report*, p. 9. In Appendix 1 of the same report, the Home Office provided supplementary evidence on access to sanitation at pp. 18–19.

maintain the minimum standard required if, as soon as a new prison is finished, it is filled to its planned capacity. Over the last few years the continually escalating prison population seems to have taken on a momentum of its own. Several attempts have been made to limit the numbers sent to prison by the provision of further alternatives to custody, notably suspended sentences, community service orders, and probation orders with special conditions,[30] in addition to such devices as parole and partly suspended sentences which have the result of shortening the total period of time spent in custody.[31] While these must have had some effect in slowing down the rate of growth, the net effect has still been steadily upwards. When combined with scepticism about the utility and value of imprisonment as a punishment, this trend has led to demands for a radical change of direction towards deliberate policies aimed at reduction. In his book, *Prisons and the Process of Justice*, Andrew Rutherford, chairman of the Howard League for Penal Reform, argues:

More is involved than curtailing the pressures which favour growth. Expansionism must be replaced by a reductionist policy that endorses three basic steps. The first step is acknowledgement that policy choices are available, as is demonstrated by developments in countries such as Japan and the Netherlands.[32] The second step requires targeting new low levels for prison usage and population size. The third step is mustering the political will to pursue the reductionist course across difficult terrain.[33]

[30] One of the most important special conditions is that, since the Criminal Justice Act 1982 (Schedule 11, S. 4B), the courts have had the power to make a probation order requiring an offender to attend a day centre run by the probation service.

[31] See A. E. Bottoms, 'Limiting Prison Use: Experience in England and Wales', in J. van Dijk, C. Haffmans, F. Ruter, J. Schulte, and S. Stolwyk (eds.), *Criminal Law in Action: An Overview of Current Issues in Western Societies*, Gouda Quint, Arnhem, 1986, pp. 293–314. Anthony Bottoms is the present holder of the Wolfson Chair in Criminology and is also Director of the Institute of Criminology at Cambridge.

[32] For a detailed review of the development of modern Dutch penal policy, see D. Downes, 'The Origins and Consequences of Dutch Penal Policy since 1945', *British Journal of Criminology*, 22 (1982), 325–57. More recently the traditional liberalism of Dutch penal policy has come under strain owing to increasing rates of crime and greater use of imprisonment in Holland. (See L. Blom-Cooper, 'A Freedom-loving Nation is Forced to Think Again', *Guardian*, 12 Sept. 1986.)

[33] A. Rutherford, *Prisons and the Process of Justice*, Oxford University Press, 1986, p. 17.

Rutherford is right to refer to political will, since the use that is made of imprisonment does not simply, or even mainly, turn on the existence of alternative non-custodial penalties, or on the size of the prison estate, but rather on more deeply rooted attitudes towards punishment. So long as imprisonment endures as a cornerstone of the penal system, there will be pressures on judges and magistrates to make use of it. Each sentencing decision is an individual response by the court to the facts of the case before it, the offender's past record, and his future prospects.[34] The reasons for sending an offender to prison are not uniform and are unrelated to the capacity of the prisons to receive him. It is futile to dismiss considered sentencing decisions taken daily in the courts of justice as the inevitable consequence of traditional thinking and blinkered judicial outlook. Nor is any Home Secretary likely to tell the courts they must not send people to prison because the prisons are too full.[35] Nevertheless, it is important for the reductionist case to be argued and developed into a distinctive strand in the continuing debate on penal policy. Not all of those who take a close interest in penal matters will be convinced, but the sound of critical voices insisting that imprisonment should be used only as a last resort, always questioning and pressing for cogent justifications, is conducive to a more broadly based, more knowledgeable and open process of policy formation. In the ringing words of the May Committee, 'Closed institutions above all require open, well-informed discussion.'[36]

Acceptance of the legitimacy of the reductionist position, the right to be heard and listened to with respect, does not, however, imply agreement with its conclusions. Attractive as it may appear intellectually to impose a ceiling, whether by a moratorium on all new prison construction,[37] by a phased programme of prison

[34] For a discussion of custodial sentencing and the principle of parsimony, see A. Ashworth, *Sentencing and Penal Policy*, Weidenfeld & Nicolson, London, 1983, pp. 318–78.

[35] See Rt Hon. Douglas Hurd, *Parl. Debates*, HC, 97 (6th ser.), col. 42, 6 May 1986.

[36] *Committee of Inquiry into the United Kingdom Prison Services, Report*, p. 3.

[37] As advocated by the Howard League for Penal Reform in 1981 in a paper entitled *No More Prison Building*, which had been presented in an earlier version to the House of Commons Home Affairs Committee.

closures, or by a combination of the two,[38] the result would be to worsen a problem that is already chronic. Rutherford's programme for a substantial reduction in the capacity of the prison system lacks nothing in boldness; nor does he hesitate to set numerical targets: 'The reductionist target for the early 1990's should not be around 52,000 as planned by the Home Office, but 22,000, or in terms of the prison population rate per 100,000 inhabitants, not 110 but in the region of 35.'[39]

The trouble with arguments of this kind is that they are free-floating, unaffected by the winds and tides of opinion which determine the political climate. The reality is that public and political demands (or what have been perceived as such) for more and longer sentences have been matched by the judicial response so unequivocally demonstrated in the tables earlier in this chapter. There seems good reason to believe that these pressures will continue irrespective of whether the Government decides to proceed with its prison building programme, to speed it up or slow it down, or to abandon it altogether. Judges are well aware that the prisons are overcrowded, and that knowledge may have some marginal effect on the use of custody in sentencing offenders. They are also aware, for the most part, that, in the blunt words used by the Lord Chief Justice in the course of a debate in the House of Lords, 'prison never did anyone any good'.[40] Yet they do not see it as their task to adjust the supply of sentenced prisoners to fit the custodial facilities available for their reception. In these circumstances, a decision to halt the building programme and/or to close some existing prisons so as to reduce capacity is highly unlikely to have any beneficial effect on the size of prison population and could well lead to disaster. Maybe a sustained educational campaign, aimed at sentencers and policy-makers as well as the general public, would ameliorate the political climate. But until there is evidence of a change in attitude, and the existence of sufficiently firm public and parliamentary support, it is improbable that any Government, irrespective of party, will follow the reductionist course.

[38] Rutherford, *Prisons and the Process of Justice*, p. 176. Rutherford proposed that the physical capacity of the prison system should be reduced by about 45%, being the equivalent of some 18,000 places.

[39] Ibid., p. 174.

[40] *Parl. Debates*, HL, 428 (5th ser.), col. 987, 24 Mar. 1982.

If we inspect the structure more closely, we see that the central plank upholding the reductionist platform is that increases in prison capacity inevitably result in increases in the number of those committed to custody. Yet this was not the experience in Britain between the wars, when the availability of prison places exceeded the number of prisoners sentenced to imprisonment. Nor does the existence of spare capacity necessarily attract custodial sentences to fill it: there is no sign that the courts are rushing to fill the current vacancies in the detention centres. It is sometimes claimed that evidence for the proposition can be found in the United States in a substantial research project undertaken for the Department of Justice in the 1970s which looked at the effects of every prison built in the USA between 1955 and 1976.[41] The researchers found that within five years of the opening of each new prison its population had exceeded its accommodation by at least 30%. But the report's conclusion was more cautious than this finding might suggest. The authors emphasize that no causal link has been demonstrated between prison capacity and population, saying merely that if nothing is done to control the size of the population it may tend to fill existing capacity.[42] They do not suggest that new prison places will always find new occupants independently of a whole range of external circumstances.

The stated aim of the current building programme is to reduce or eliminate overcrowding by equating total available prison places and average projected population levels by the early 1990s. This objective may or may not be achieved, but at least it has the merit of being a clear-cut aim, albeit one that carries political risks if it is not attained. Important as it is, adding to the overall stock of accommodation is not the only desirable end. New places are required to permit sub-standard accommodation to be temporarily vacated for modernization and refurbishment, or, if this is impracticable or uneconomic, to be closed and

---

[41] National Institute of Justice, *American Prisons and Jails*, vol. i, US Department of Justice, Washington, DC, 1980, p. 94.
[42] James Q. Wilson rejects the findings as providing no valid support for the capacity model. He states that an investigation by the Panel on Sentencing Research of the National Research Council concluded that the findings were seriously in error. *Thinking About Crime* (rev. edn), Basic Books, New York, 1983, p. 161.

disposed of. There is no automatic link between the scale of the building programme and the operational capacity of the prison estate. The extent to which it may in future be possible to use the new accommodation to replace rather than supplement existing facilities will depend on population pressures at the time. This also bears on the central issue of prison conditions, since at present the need to decant inmates while modernization takes place both compounds the problem of overcrowding and delays the provision of such urgently needed improvements as integral sanitation. In this way, the physical capacity of the prison system is directly related to the conditions within it.

The politics of prison building lie more in successive Governments recognizing, but finding it hard to fulfil, an administrative need inhibited by a reluctant public than in differences between the main political parties. Thus James Callaghan, when delivering a Home Office Bicentenary Lecture in 1982, spoke in terms untainted by party ideology:

Inability to build new prisons has been a continual headache for every Home Secretary, and the fact that there was a period of relative quietness between the two world wars only concealed and compounded the problem.... Whenever we discussed levels of public expenditure in Cabinet, I never had any doubt that, if the choice lay between spending on a new school or a new prison, it would be the Secretary for Education who would win the prize and the poor Home Secretary who would lose out.

The neglect of governments is there for all to see. Fresh funds are so often only forthcoming in response to a crisis that I was pleased to secure a modest £20 million building programme, even though it was still not adequate at that time.... Later events falsified at least my financial hopes and, looking back, they now appear to have been overoptimistic. If we had maintained those plans, I wonder, would we have done better in avoiding some of the recurrent troubles which have plagued our prisons in the last decade? I am certain that we need to spend more money on our penal institutions.[43]

It has taken a monumental effort to get the present prison building programme under way and it deserves all-party support. The task now is to prevent slippage and cutbacks; to see the new buildings and improvements through to a conclusion.

[43] J. Callaghan, *The Home Office: Perspectives on Policy and Administration*, Royal Institute of Public Administration, London, 1983, p. 16.

V

The massive investment in new prisons has also been criticized on the grounds of draining resources away from the operational needs of the prison system, in particular, education[44] and training. Anyone who visits the prisons at all regularly will have been struck by the sight of so many empty workshops or classrooms. The facilities and instructors appear to be there, but not the prisoners. There are organizational as much as financial reasons why this should be so; and personal observation is supported by the official *Report on the Work of the Prison Department 1984/85*, which commented on the under-utilization of workshops, plant, and equipment, stating that fundamental changes were necessary to arrest the decline of prison industries and make more effective provision for inmate employment.[45] In view of the growth in the size of the prison population, it is disheartening to find that the number of industrial hours worked in the prisons has dropped by more than 50% since the early 1970s, and that in March 1986 one in three workshop places were unfilled.[46] Measured in terms of inmate hours worked, this represented a reduction to about $8\frac{1}{2}$ million hours in 1985/6, compared with about 21 million a decade earlier.[47]

Restrictive staff practices and disputes about manning (with implications for overtime pay) have played a part, while some prisoners may contrast the traditional emphasis on industrial training with the limited job opportunities outside. Nevertheless, enough well motivated prisoners either already possessed or learned sufficient craft skills to form the work-force at a number of industrial prisons, notably Coldingley, which is run on factory lines. Elsewhere, full-time vocational training courses allow selected prisoners to obtain a qualification. But the general picture is one of large numbers of inmates without previous

[44] The NACRO *Annual Report 1984/85* stated that money for education was being reduced from 2.3% of the prison budget in 1979/80 to 1.85% in 1984/5. A senior Prison Department official was quoted as having told the House of Commons Select Committee on Education that there had been some reduction because of the need to pay for new prisons and more prison officers. p. 6.

[45] *Report on the Work of the Prison Department 1984/85* (Cmnd. 9699), HMSO, London, 1985, p. 45.

[46] *Parl. Debates*, HC, 94 (6th ser.), col. 408, 25 Mar. 1986.

[47] *Report on the Work of the Prison Department 1984/85*, p. 45.

experience of regular employment, sometimes through lack of opportunity but also because of delinquent life-styles, low standards of intelligence, or other inadequacies or disabilities, whiling away their time in custody with little in the way of treatment and training to stretch their minds or develop their outlook. For too many, the designation 'training prison', still used to describe those prisons containing lower-security category inmates serving medium to long sentences, is a complete misnomer. A remark made by a prisoner to a visiting newspaper reporter rings true of the prison system as a whole: 'There's nothing to do; so we just do nothing.'[48]

More can and must be done to break out of the vicious circle of apathy, boredom, anxiety, and despair that is prompted by lack of work, training, or even physical exercise. Virtually any form of occupation must stand a better chance of sustaining, sometimes even enhancing, the healthier aspects of a prisoner's personality than idleness or meaningless drudgery. Work in the kitchen or cleaning is dull and uninspiring, and industrial training, where it exists at all, may seem to be of limited relevance to a prisoner's life on release. But training can be broadened into social skills and the constructive use of leisure. One Scottish prison is reported to have an aviary and to be breeding fish for Third World countries.[49] Many more prison establishments could follow the shining example of Preston, one of the oldest prisons in the country and set in entirely urban surroundings, which has created an exceptional garden inside the perimeter wall with the assistance of the inmates.[50] Staff and prisoners are brought together in a common endeavour, their work in the grounds improving the amenities and appearance of the normally bleak prison setting, as well as cultivating a spirit of achievement and pride of place. For people who often have attained little in their lives, any achievement is worth-while, particularly one involving experience that can be put to good use outside. Education at various levels is similarly important, and it is encouraging to find that some of those who have completed

[48] The *Independent*, 24 Nov. 1986.
[49] The *Observer*, 7 Dec. 1986.
[50] HMP Preston twice won the Windlesham Trophy for the best kept prison garden, organized with the assistance of the Royal Horticultural Society, in the first three years of the award, 1984–6. Liverpool Prison's garden was featured at regular intervals by Granada Television.

educational courses in prison continue with their education or training at a higher level on release. Remedial English teaching to prisoners from Asian or other non-English-speaking backgrounds can also be a lasting benefit. No attempt has yet been made to quantify the effects of the more imaginative schemes in terms of future re-offending, and so at present prison managements have to be guided by optimism and common sense rather than by any conclusive evidence.

The declining opportunities for work are an authentic sign of the changing prison ethos. The causes are to be found in the way prisons are run and the balance of forces within and outside them, as was shown in the sorry story of the prison industry losses. The origins go back to the 1970s, when the pressure to provide work for prisoners, a highly commendable aim, coincided with another aim, equally valid: to seek the best possible financial return on the money invested in plant, components, and equipment. The resulting confusion of objectives, penological and commercial, was never satisfactorily resolved. A policy of producing goods for sale was introduced without adequate assessment of potential markets or effective controls on predicted or monitored levels of loss. The eventual result was a public scandal. Losses to the taxpayer amounted to £17.3 million written off in 1984/5 and bad debts of a further £1 million in respect of prison industry contracts. The Comptroller and Auditor-General warned that the sums written off could rise to £20 million, with an additional £5 million for excessive stocks and equipment.[51] The Home Office was censured by the House of Commons Committee of Public Accounts for a catalogue of 'inefficiency and bad management and delay', the Committee concluding, on the basis of the Comptroller and Auditor-General's report and additional information supplied to it by the Permanent Under-Secretary as Accounting Officer, that the Department had 'fallen far below acceptable standards of public administration'.[52]

Worse even than maladministration was the suspicion of fraud and corruption, which led to criminal proceedings being taken against certain civil servants and contractors. The charges were

[51] House of Commons Committee of Public Accounts, *Twenty-Sixth Report*, Prison Industry Losses, Session 1985/6 (HC, 160), HMSO, 1986, p. v.
[52] Ibid., pp. vi–vii.

not proved in court, although internal disciplinary measures were subsequently taken by the Home Office. This chapter is now closed; the objectives and organization of prison industries have been overhauled and a new management installed, made answerable to a Board including two outside non-executive members. Nevertheless, a setback of this magnitude must leave a mark, not only on all the officials involved, but also on the impetus towards constructive work and training in the prisons. It would be a tragedy if the aftermath were to be an acceleration of the emptiness of prison life caused by failures of organization.

The Home Office has other problems, less dramatic but more deep-seated, in running the prisons. There is, first of all, a yawning gulf between what the national bureaucracy wants to do and what actually happens in the prison establishments throughout the country. Monolithic as the Prison Department may appear to be to those inside, the reality, as noted by the May Committee[53] and explored in a detailed survey of staff attitudes in the prison service conducted by the Office of Population Censuses and Surveys (OPCS),[54] is otherwise. The OPCS report documented a widespread sense of parochialism and homogeneity binding staffs together in distinct and separate communities, making it difficult for the aims and methods of the central management of the service, located within the Home Office in London, to be appreciated and carried out by the uniformed staff and other grades in the prisons. But it is more than individuality that divides the staff on the ground from the policy-makers: routine pressures bear heavily on many parts of the prison system.

Twenty years ago, the Mountbatten Inquiry into prison security, set up in 1966 after a series of spectacular escapes culminating in that of the spy George Blake, concluded that there was no really secure prison in existence anywhere in the country.[55] To remedy this state of affairs, Mountbatten recom-

[53] *Committee of Inquiry into the United Kingdom Prison Services, Report*, pp. 22–4, 74–90.

[54] A. Marsh, J. Dobbs, J. Monk, with A. White, *Staff Attitudes in the Prison Service*, an inquiry carried out on behalf of the Home Office. Office of Population Censuses and Surveys, Social Survey Division, HMSO, London, 1985.

[55] *Report of the Inquiry into Prison Escapes and Security* by Admiral of the Fleet Earl Mountbatten of Burma (Cmnd. 3175), HMSO, London, 1966, p. 4.

mended a new system of categorizing inmates according to the degree of threat to the public which their escape would pose. Other organizational proposals included the appointment of an Inspector-General to be the recognizable head of the prison service, and a strengthening of the inadequate arrangements for inspecting the prisons. The form of inspection, and the practicality of making it an independent function distinct from the Prison Department, continued to be vigorously debated for several years to come.[56] But over the prison system as a whole, there was no doubt that a greatly enhanced emphasis on security followed. Its effect was spread more widely as a result of a decision by the Government of the day to disperse the category A high-risk prisoners across a number of prisons rather than concentrating them in a single maximum-security prison as favoured by Mountbatten.[57] Consequently, seven of the existing training prisons had their perimeter security upgraded and their regimes overhauled to become the dispersal prisons. It is in these establishments that the most dangerous and disruptive prisoners have been held ever since. Other inmates in lesser-security categories are also allocated to the dispersal prisons. While the higher standards of perimeter security and control within these prisons have combined to prevent a recurrence of the sensational escapes from custody which were so damaging to

[56] A separate inspectorate of prisons was introduced in 1969, but it was initially placed within the Prison Department and its reports were not made public. In 1979 the May Committee, after reviewing what it described as 'the most difficult organization issue' which it faced, recommended that there should be a system of inspection distanced as far as was practical from the Prison Department, and that reports of the inspections should be published. This recommendation led to the establishment of the post of HM Chief Inspector of Prisons, appointed by the Crown on the advice of the Home Secretary. The Chief Inspector reports directly to the Home Secretary and has his own staff. The terms of reference and constitution of the inspectorate, commonly known as its charter, are contained in an appendix to the first *Report of HM Chief Inspector of Prisons for England and Wales, 1981*, published by HMSO in March 1982 (Cmnd. 8532). The inspectorate was placed on a statutory footing in S. 57 of the Criminal Justice Act 1982. For the historical background, see E. Stockdale, 'A Short History of Prison Inspection in England', *British Journal of Criminology,* 23 (1983), 209–28.

[57] This followed a recommendation by a sub-committee, chaired by Professor Leon Radzinowicz, of the Advisory Council on the Penal System, which came down firmly against the proposal to concentrate high-risk prisoners in a single maximum-security establishment. *The Regime for Long-term Prisoners in Conditions of Maximum Security,* HMSO, London, 1968.

public confidence, the price has been mounting inmate unrest. Prisons have always been volatile—even explosive—communities, and the friction between differing types of inmates in the dispersal prisons, reacting against stricter regimes, has bred indiscipline and violence. Riots have broken out, complete wings have been ravaged, and staff abused, assaulted, or held hostage. No one should underestimate the constant danger to which prison officers are exposed in such circumstances.

It is necessary to know and understand some of this background to get to the root causes of the staff discontent which, together with inmate unrest and overcrowding, has been one of the most notable features of prison life over the last two decades. Frustration was experienced by prison officers, who felt that, in the more autonomous prisons of the 1960s, they were expected to support and implement ideas about which they had not been consulted and which did not conform to any recognizable national strategy. In a bitter comment, the Prison Officers' Association complained:

Prison Officers found themselves foisted by decisions they were never consulted about. Governors came and went, and local regimes were run to suit the whims of individuals who believed their career would be enhanced the more they conformed to the demands of the inmate population.[58]

It is perhaps natural that prison officers should feel aggrieved by what some of them regard as the disproportionate attention paid to the rights of the prisoners rather than the requirements of the staff. The depth of resentment was reflected in the survey of staff attitudes referred to above. This indicated that officers felt they got too little support, that the prison authorities were too frightened of public opinion, and that society cared more for the welfare of prisoners than for the welfare of prison staff. Prison officers, particularly the younger ones, said they felt isolated from senior management. Overall staff attitudes tended to be cynical and jaded, the chosen adjective being 'jaundiced'. Prison officers were found to have little pride in their jobs, many preferring to keep their occupations to themselves rather than tell people outside that they worked in the prison service. They

[58] Prison Officers' Association, *Financial Management Initiatives*, July 1985, p. 2.

were sensitive, feeling that society looked down on them or, worse still, saw them as bullies, hard and unfeeling, with an unmerited reputation for callousness. Negative public attitudes, some thought, spilled over to affect their wives and children.[59] Factors of this sort, accentuated by the geographic remoteness of certain prisons, including most of the dispersal prisons, help to explain the inward-looking quality of the prison service, with prison officers mixing socially more among themselves when off-duty, by choice as well as convenience, than with the population at large.

VI

To isolation, sensitivity, and low morale must be added grievances over pay, allowances, manning, and patterns of work as the underlying causes of the escalating conflict between staff and management in the prison service. Inflexible and inefficient practices have grown up which, in the estimate of a published report of a joint study by the Prison Department and management consultants, have resulted in the prison service operating at some 15–20% below what should be its optimum capacity.[60] The way prison officers are paid is complicated, with a system of allowances and other variable payments supplementing a basically low rate of pay. Thus, remuneration in the prison service has become dominated by overtime: the Home Office stating in 1986 that, on average, 30% of prison officers' pay consisted of overtime and in many cases the percentage was even higher.[61] As a result, some officers were working in excess of twenty to thirty hours' overtime a week, a situation described by the Home Secretary as being not good for them or for management, and one that could not be prolonged indefinitely.[62] There can be no doubt that the present arrangements are a bar to

---

[59] OPCS, *Staff Attitudes in the Prison Service*, summary of main findings, pp. 96–102.

[60] HM Prison Service, *Study of Prison Officers' Complementing and Shift Systems*, joint study by Prison Department and PA Management Consultants, 1986. Volume I: *Report*.

[61] *A Fresh Start* produced for the Home Office by the Central Office of Information, London, July 1986, p. 7.

[62] *Parl. Debates*, HC, 97 (6th ser.), col. 44, 6 May 1986.

efficient working, with staff depending excessively on total earnings boosted by overtime, which cannot be guaranteed.

Reforms are now urgently needed to bring working practices into line with operational requirements, providing managements with flexible systems better able to meet the special character of each establishment. The intention is that overtime working, the heart of the problem, should cease, to be replaced by a system of contracted hours and higher rates of basic pay. The Prison Department accepted that pay levels under the new arrangements must take account of current levels of average earnings, reflecting a balance between staff numbers, hours worked, and pay in order to ensure that the prison service can carry out the work it has to do within the overall finance available to it.[63]

None of this will be achieved easily. Prison officers are unionized, and their union, the Prison Officers' Association (POA), has been ready to take or support industrial action in what it sees as the protection of its members' interests. Strained relations between staff and management are no new occurrence in the prison service: the 1950s saw the first serious outbreak of industrial unrest, followed by a calmer period in the 1960s and early 1970s.[64] Then, however, growing inmate unrest and violence, erupting in serious riots at two of the dispersal prisons (Albany and Gartree) in the autumn of 1972, led the POA to threaten a national strike unless levels of staff were increased. Mounting concern over security and manning coincided with greater militancy in the pursuit of disputes over pay and conditions. In this, the prison service may to some extent have been reacting to the turbulent industrial climate outside. In any event, at Cardiff in 1975 the inevitable happened, and for the first time since 1919 one of HM prisons experienced an almost total withdrawal of labour for the best part of a day.

Following a decision by the POA national executive in September 1975, the power to take industrial action in furtherance of disputes between management and staff passed increasingly to the local branches of the union. The deterioration in industrial relations which resulted was so marked that it led, in November 1978, to the establishment of a Departmental

---

[63] *A Fresh Start*, pp. 7–8.
[64] *Committee of Inquiry into the United Kingdom Prison Services, Report*, p. 233.

Committee, under the chairmanship of a High Court judge, Mr Justice May,[65] to inquire into the state of the prison services throughout the UK. Prisons in Scotland and Northern Ireland as well as in England and Wales were included in the review. In addition to wide-ranging terms of reference relating to 'the size and nature of the prison population and the capacity of the prison services to accommodate it', the Committee was requested to inquire into 'the adequacy, availability, management and use of resources', as well as making recommendations on the remuneration and conditions of service of prison officers, governors, and other grades.[66] The Home Secretary, Merlyn Rees, stressed the urgency of the task, originally expressing the hope that the Committee would report in six months. In the event, the thoroughness of the investigation ruled this out, and the Committee's Report, submitted to a new Home Secretary (William Whitelaw) and published within twelve months of its appointment, was a considerable accomplishment.

The May Report is the most comprehensive account available on the state of British prisons—their populations, organization, and staffing. Set up by a government of one complexion and reporting to that of another, it is free from political bias and is an invaluable repository of factual information, much of it not previously accessible to the general public. On the emotive subject of industrial relations, for instance, the Report dispassionately recorded that, in the six months between its establishment and the end of May 1979, industrial action took place on some 31 occasions at 25 different establishments, in addition to the general civil service industrial action over pay in April, which affected regimes at 105 establishments.[67] Another well-informed independent commentator, Professor Michael Zander, took up the same theme in a Home Office lecture. After referring to negative feelings about managements throughout the prison system and to defensive management attitudes, he went on:

the state of industrial relations in the prisons verged on open hostilities, or even anarchy. Industrial action taken by staff included refusal to allow solicitors, police officers and probation officers into prisons; preventing

[65] Sir John May was a judge of the High Court, Queen's Bench Division, 1972–82, and Lord Justice of Appeal since 1982.
[66] *Committee of Inquiry into the United Kingdom Prison Services, Report,* p. iii.
[67] Ibid., p. 235.

the appearance of prisoners in court; refusing to admit those committed to prison by the courts; late unlocking of prisoners or refusal to unlock them at all; and restricting the movement of supplies into and out of prisons. Officers had prevented classes, exercise, visits and association, and on occasion had caused the inmates to remain locked in their cells for 23 hours a day. On the other hand, the Home Office appeared to have opted out of the handling of many of the disputes. According to the May Committee, the Prison Officers' Association had formed the view that it would get its way through tough action and that officers would be able to take industrial action with impunity.[68]

The move towards fundamental organizational change has been gradually gathering pace. In May 1985 a clear indication of the Government's resolve was given by Leon Brittan in a speech to the POA National Conference in Portsmouth. After announcing some of the immediate measures being taken to cope with the rapid rise in population in the early months of the year, and the additional places and staff that had resulted, the Home Secretary said:

The treatment I have outlined is exceptional in the public service in the present economic climate. If we are to justify this growth, I need to be able to assure my Cabinet colleagues and the public that the Prison Service is making the best use of the resources it has been given. Management has embarked on a programme designed to allow me to do this. I look to you to support it. I realize that it means some changes in hallowed work practices. But increased efficiency must go hand in hand with increased resources.

As so often happens when the protection of established interests collides with inadequate funds to meet the expectations of staff, the path of negotiation did not run smooth. Before twelve months was out another Home Secretary, Douglas Hurd, was reporting to a shocked House of Commons that widespread disruption had occurred in many parts of the prison system after the POA had instructed its members to withdraw co-operation with management and institute an overtime ban in April 1986. Major incidents, including violence and arson, were reported at four prisons and one youth custody centre, and lesser incidents in twelve other establishments. Although fortunately there were

[68] Michael Zander is Professor of Law at the London School of Economics. His bicentenary lecture, in the same series as that given by Mr Callaghan (n. 43 above), was published in *The Home Office: Perspectives on Policy and Administration*, Royal Institute of Public Administration, London, p. 66.

no serious injuries, fifty or so inmates made bids to escape, and a week later thirteen were still unlawfully at large. Prison buildings were damaged in a large number of establishments, seriously so at four, while at one the damage was so extensive that the entire prison had temporarily to be vacated. In all, about 800 prison places were put out of action.[69]

Three months later, in July, the Home Secretary announced a package of proposals covering new working arrangements for prison officers and changes in management and pay systems. The proposals, described as *A Fresh Start*, were detailed and supported by the findings of a study by a joint team of prison service officials and management consultants on shift and complementing systems. A summary was issued to all members of the prison service in England and Wales in leaflet form backed up with a video. In November 1986 a formal offer was put to the National Executive Committee of the POA.[70] It is too early to predict the outcome. Entrenched interests will not yield easily, and criticism, from special interest groups as well as political opponents, can be expected to concentrate on the negotiating tactics, the financial provisions, and the short-term effects on the service. But anyone with the true interests of the prisons and prisoners at heart must find it hard to challenge the objectives. It was, after all, a Labour MP, at the time chairman of the Parliamentary All-Party Penal Affairs Group, who told the House of Commons, in the aftermath of the devastation in the prisons in April 1986:

No one, including the POA, prison officers, the prison service and the Labour movement, will do anyone any favours by pretending that there are not abuses of the overtime system and that there are not restrictive practices which should be eliminated. There are several abuses of working practices and the overtime system, and these relate to escort duties, the shift system, weekend workings at overtime rates and the insistence of prison officers that they escort prisoners to education classes and workshops. There are abuses, and they need to be dealt with effectively. . . .[71]

[69] *Parl. Debates,* HC, 97 (6th ser.), col. 46, 6 May 1986.

[70] The terms of the offer were published and circulated to members of the prison service in a further leaflet entitled *A Fresh Start—What it means for you* (produced for the Home Office by the Central Office of Information, London, November 1986).

[71] Robert Kilroy-Silk, MP: *Parl. Debates,* HC, 97 (6th ser.), col. 65, 6 May 1986.

VII

There is no single solution to the manifold problems presented by prisons and imprisonment. While there have been some steps forward, they have been accompanied, as we have seen, by steps back. The entire prison system—the buildings, the staff, the prisoners, and the regimes—is threatened by overcrowding. The scope for dramatic easement which existed in the early years of the century, and which was exploited by Churchill with such flair, no longer exists. Then, the large numbers of adults imprisoned for non-payment of fines and drunkenness, together with young offenders in prison for very short periods for trivial offences, provided an opportunity for reduction which did not inflame public opinion. Today, the only hope of progress lies in pursuing a number of complementary policies with equal vigour. On the demand side, overcrowding can be countered to some extent by ensuring that the courts have an adequate range of non-custodial penalties at their disposal and are encouraged to make use of them to the full. Ministers are constrained by the concept of judicial sentencing discretion to exhortation and periodic legislative intervention. The most profound change brought about by legislation, originally in 1967 but greatly extended later, is the facility to shorten the sentences imposed by the courts by means of parole. Now running at about 14,000 early releases on licence in a year, and thus a highly significant feature of the penal system, parole was not originally conceived as a mechanism for limiting the size of the prison population. Nor are parole decisions on individual cases influenced by population pressures; the way they are taken, and the criteria employed, are discussed in the following chapter.

As a long stop, there is a power of executive release, originally conferred on the Home Secretary by Parliament in emergency legislation resulting from industrial action in 1980, when prison officers refused to admit any more prisoners pending the resolution of a dispute over meal breaks. The power was re-enacted in the Criminal Justice Act 1982 and enables the Home Secretary, subject to affirmative resolution of both Houses of Parliament, to order the release of certain categories of prisoner up to six months before they would otherwise be released if 'he is satisfied that it is necessary to do so in order to make the best use of the places available for detention'.[72] Despite being urged to activate this provision, successive Home Office Ministers have made it plain, in Parliament and outside, that they regard it as an emergency power to be used only in the event of a breakdown in the prison system.[73] They are not prepared to contemplate its use simply to create a breathing space. In principle, this must be correct; it is for the courts to pass the sentence and for the executive to see that the proper proportion of it is served unless, as in the case of parole, Parliament has approved a specific scheme holding out the promise of a greater degree of protection for the public in the future. This is what the public expects, and any blurring of the dividing lines, seemingly in the interests of administrative expediency, would risk undermining public confidence, already precarious enough, in the operation of the penal system at one of its most sensitive points.

A more promising line of approach is to explore alternative ways of handling those prisoners who have not been sentenced at all, but are held in custody awaiting trial. By December 1986 the prison population had climbed to 48,342,[74] the highest number ever recorded in modern times. Included in this total were over 10,000 remand prisoners. Sheer pressure of numbers, rather than any decisions on principle, may bring to a head the issue of privatization within the prison system. While there are strong arguments against the state handing over to private enterprise responsibility for the custody of convicted and sentenced prisoners, these objections do not apply with equal force to those individuals, now numbering one in five of the prison population, who have been charged with a criminal offence and are awaiting trial. As mentioned earlier in this chapter, a relatively small minority will be acquitted, and a higher proportion will be

---

[72] Criminal Justice Act 1982, S. 32(1). The power does not extend to those serving sentences of life imprisonment or to prisoners who have been convicted and sentenced as a consequence of committing an excluded offence. These offences, largely of violence to the person, drug trafficking, or terrorism, are listed in the first schedule to the Act. Attempting or conspiring to commit such an offence, and aiding or abetting, counselling, procuring, or inciting the commission of such an offence, are brought within the ambit of the exclusions by S. 32(1) (b) ii–iv.

[73] See Patrick Mayhew, the Minister of State at the Home Office and later Solicitor General, *Parl. Debates* (HC) SC (1981–2), 1, col. 421, 4 Mar. 1982.

[74] Including those held in police cells.

convicted and sentenced to non-custodial penalties.[75] All are entitled to be treated as innocent until proved guilty, and there is much to be said for separating more clearly those held on remand from sentenced prisoners. The most effective way to distinguish the two categories would ·be to detain remand prisoners in separate establishments. Some remand centres already exist, but they are prisons by another name and staffed by prison officers. Moreover, the flow of accused persons who have been remanded in custody by the courts is on such a scale that large numbers of remand prisoners spill over into the local prisons.

Punishment is the touchstone that should mark out sentenced prisoners from those who are awaiting trial on remand. Convicted offenders who are sentenced to a term of imprisonment for their crimes are being punished; untried persons are not. In their case, the justification for detention is entirely preventive: to prevent the possibility of further offences being committed; to prevent witnesses from being tampered with; and to prevent accused persons from disappearing and failing to come to court to answer the charges against them. These are the reasons why the courts are empowered to deprive of their liberty for a temporary period those charged with a criminal offence if satisfied that bail cannot reasonably be granted. It is quite wrong that the conditions in which convicted prisoners are held in custody for punishment should be so similar in their essentials to those in which untried persons are detained.

Remand centres outside the prison service, employing agency staff and licensed and supervised by the Home Office, would at once relieve the pressure on the prisons and signify the special status of the inmates. Escorting and transporting remand prisoners to and from the courts is labour-intensive and occupies many hours of prison officers' time which could be put to more constructive use at the prisons. New-style remand centres, leased from the Government and subject to inspection, would need to operate according to strict requirements as to security and the

[75] Of all persons charged and remanded in custody by magistrates (47,100 in 1985) some 2,400, or 5%, were acquitted by the court and a further 20% given a non-custodial sentence. Of the 6,200 people in 1985 remanded in custody during their trial at the Crown Court, 6% were acquitted and a further 11% given a non-custodial sentence.

handling of inmates. Special arrangements could be made, on application to the court, for high-risk prisoners to be remanded only to a prison. Precedents for the use of contracted agencies already exist in the immigration service, where employees of a private security organization are responsible for the custody and escort of illegal immigrants and others who are refused entry to the United Kingdom or are required to leave after entry. In 1985, about 7,000 people were held overnight or for longer periods in detention accommodation for which the immigration service was responsible. A further 1,000 were received into prison under Immigration Act powers. Ideological opposition to privatization should not prevent a move in this direction, which would do much to ease the crisis in the prisons without breaching any indispensable principle.

Whatever the arrangements made for their detention, the number of untried prisoners held on remand at any one time could be reduced if they did not have to wait so long for trial. This matter is outside the hands of the Prison Department, but the need to minimize delays, speeding up the progress of business through the courts, is widely recognized. One avenue is the setting of time limits, as authorized by the Prosecution of Offences Act 1985, so that the period spent in custody by persons charged with criminal offences but refused bail is no longer unlimited. There is some prospect of relief here, and it is intended that statutory time limits will be introduced in three police areas[76] from 1 April 1987.

On the supply side, government has a freer hand to increase the overall number of prison places, subject to the availability of financial resources and physical restrictions, such as planning consents and local opposition, which can impede the acquisition of sites and new building. Tactical management can also be instrumental in reducing overcrowding. If more can be discovered about the response of prisoners to different regimes, so that a closer fit can be obtained between the type and distribution of prison places and the varied categories of prisoners to be accommodated, the prison estate might then be utilized more intensively.

All of these objectives need to be pursued simultaneously, and

---

[76] Avon and Somerset, Kent, and the West Midlands.

with a sense of urgency. It is regrettable but inevitable that more new prisons and remand centres are necessary, although, as has been argued, the case for the building programme rests not just on the provision of additional places, but on the paramount need to improve standards and regimes across the prison system as a whole, particularly in the local prisons. It is simply unacceptable to continue to tolerate without protest conditions that allow for so little respect to human dignity. Reasons of this kind moved one of the founding fathers of English criminology, Sir Leon Radzinowicz, to declare flatly and without qualification that 'building prisons today is a measure of penal reform'.[77]

VIII

I am conscious that in this chapter, unlike some of the earlier ones, I have hardly touched on the situation of the convicted offenders who make up the bulk of the prison population, concentrating instead on the institutions to which they are sent. This is deliberate, since it seems to me evident that considerations of penal policy and sentencing, as well as prison capacity and organization, are the key issues at present. But no commentary on imprisonment should conclude without coming back to the prisoners. However sparingly deprivation of liberty is used as the most severe sanction available to the courts, there will always be a supply of men and women arriving at the gates to keep the prisons in business and the public protected. How they are treated in custody—by established authority, by prison officers, and by each other—are questions that are fundamental to the day-to-day existence of many thousands of inmates. The need for security is the prerequisite for all establishments; if prisons are not secure, they fail in their primary social purpose. But, provided prisoners are accurately assessed and allocated, not all establishments will need the same levels of security. Second only to security, and compatible with it, is the duty of the state and its servants to manage the prisons in a decent and humane way. Like so much else in the field of crime and the punishment it attracts, this is much easier to proclaim than to act

[77] 'Law in Action', BBC Radio 4, 24 Jan. 1986.

upon, particularly in conditions of overcrowding which are aggravated by inflexible and inefficient working practices, inhibiting rather than encouraging the more positive aspects of imprisonment.

There can be little doubt that treatment and training has had its day, having had a good run and an honourable one, and that it is no longer in tune with the public mood, nor with the realities of everyday life in prison. If dispensable, however, something must be put in its place. The corollary of what is coming through as the dominant penal philosophy on both sides of the Atlantic—the perception of a term of imprisonment as the 'just deserts' earned by an offender as the consequence of his crime—is the policy of humane containment. This is all very well as far as it goes, and is greatly to be preferred as an aim to the unspoken alternative to inhumane containment. Yet, unfortunately, on any scale calibrated between humane and inhumane, far too many prisons, or parts of prisons, would be clustered closer to the latter end.

These are criticisms of the way prisons are run rather than of the dedication of the staff who work in them. The distinction needs to be kept in mind if the organizational blocks which for so many years have stood in the way of progress are to be dismantled. Like prison building, this too is a legitimate objective of penal reform. For many people, a career in the prison service still has an element of vocation. It is hard and demanding work, often in sub-standard buildings in grim surroundings, with a sense of tension and danger never very far away. Prison officers have a thankless job, and it is not surprising that they feel unappreciated by the wider public. In the attitude survey already quoted,[78] more than 3,000 staff representing all grades were interviewed, and their desire for a more creative and more fulfilling role came through strongly. Such aspirations need to be cultivated.

All those with first-hand experience of visiting prisons know that the high ideal expressed in the first of the Prison Rules—'the purpose of the training and treatment of convicted prisoners shall be to encourage and assist them to lead a good and useful life'—is far removed from the reality. Seven years ago, after

[78] OPCS, *Staff Attitudes in the Prison Service*.

seeking and receiving oral and written evidence, the May Committee came to the same conclusion, recommending that Rule 1 should be rewritten rather than scrapped completely. Their report warned that prison staffs cannot be asked to operate in a moral vacuum, and that the absence of real objectives could only lead in the end to the routine brutalization of all the participants.[79]

The emphasis on management and regimes, important as it is if long-overdue reforms are to be achieved, must not be allowed to obscure the need for guiding principles. What is now called for is a reformulation of Rule 1 along these lines: establishing a secure environment within which all staff accept a responsibility to help prisoners develop the better side of their personalities;[80] preserving self-dignity and promoting self-esteem and respect for others; minimizing to the greatest extent possible the harmful effects of incarceration; maintaining hope for the future; and preparing inmates for their eventual discharge. Such objectives transcend the sterile and inconclusive debate on whether such rehabilitation as can be achieved in the prison setting has any effect on subsequent re-offending. The proclaimed values are spiritual and enduring: they do not depend on measurable results. Many members of the prison service instinctively respond to sentiments of this tenor, being averse perhaps to anything too high-flown, but acknowledging in their consciences, and sometimes in their public statements,[81] the obligations placed on them towards those in their charge. They deserve every encouragement to act according to these lights.

[79] *Committee of Inquiry into the United Kingdom Prison Services, Report,* p. 67.

[80] In an eloquent speech in the House of Commons in 1910 Winston Churchill, when Home Secretary, spoke of 'a desire and eagerness to rehabilitate in the world of industry all those who have paid their dues in the hard coinage of punishment, tireless efforts towards the discovery of curative and regenerative processes, and an unfaltering faith that there is a treasure, if you can only find it, in the heart of every man ...' Quoted by Sir L. Radzinowicz and R. G. Hood, *A History of English Criminal Law and its Administration,* vol. 5, Stevens, London, 1986, p. 774.

[81] 'Prison Officers have made it clear that they are committed to ensuring that inmate care; inmate safety; the safety of staff and the safety of the public are immutable principles.' Prison Officers' Association, *Financial Management Initiatives,* p. 17.

8

Release on Licence: Mitigating Punishment and Protecting the Public

I

Prison is no longer the last stop in the continuum of offence–detection–charge–trial–conviction–sentence. Since 1967, legislation has permitted the early release of prisoners in England and Wales who have served a prescribed period in prison and who have satisfactorily completed a proportion of their sentence in custody (not less than one third), enabling them to serve the balance of their sentence in the community. Scotland has a similar, although separate, system. In Northern Ireland there is no parole, but 50% remission rather than one-third as elsewhere in the United Kingdom. For those prisoners who are selected for parole, after a lengthy process of reporting and review in which the protection of the public plays the largest part, the constraints of the prison walls are exchanged for the lighter restrictions and the supportive potential of a parole licence.

Parole can be seen as a pact by which the punishment imposed by the court is mitigated in exchange for the prospect of a reduction in the incidence, gravity, or frequency of future offending. This is what is meant when the Parole Board says, as it did in the opening paragraphs of its annual reports for 1984 and 1985, that the protection of the public is its primary and predominant concern.[1] Other justifications can be and have been offered for early release on licence, but this is the social compact that characterizes parole as it is generally understood today. It is this ground, more than any other, on which depends the

[1] *Report of the Parole Board 1984* (HC, 411), HMSO, London, 1985, p. 1; and *Report of the Parole Board 1985* (HC, 428), HMSO, London, 1986, p. 1.

continued public acceptance of a system which in 1985 facilitated the early release from custody of over 14,000 prisoners.[2]

The introduction of parole in the Criminal Justice Act 1967 marked a step forward in the penal system. For decades, penal reformers had been arguing that the purposes of punishment and its optimum impact on the prisoner could be achieved without the necessity for all convicted offenders to remain in custody for the whole of the term of imprisonment imposed by the courts. Another aim was to humanize the prison system by providing an inducement to prisoners to improve their characters and holding out the hope of early release. The concept of parole was not a new one: prisoners had long been allowed out temporarily to attend funerals, weddings, or other family occasions, and prison governors had discretion to agree to short periods of pre-release leave in appropriate cases. The essence of these arrangements was that the prisoner was on his honour to return and to observe any other conditions which were part of the bargain. The element of mutual trust was carried over into the parole scheme in the early days, although it has now diminished to such an extent as to be of negligible relevance.

Ever since 1901, provision for early release had existed in New York[3] and some other state jurisdictions in the United States, while a system of parole for prisoners convicted of criminal offences in the federal courts was introduced in 1930. The same current of penal reform favouring supervision and rehabilitation in the community had been running on this side of the Atlantic, although here the main thrust was towards the establishment of probation as an alternative to imprisonment. After some spasmodic and isolated initiatives at the close of the nineteenth century, the Probation of Offenders Act 1907 authorized the courts to appoint probation officers and to make orders committing offenders to the care and supervision of such officers. In 1925 a requirement was placed on the courts to appoint probation officers, an obligation that remains today.[4]

[2] See *Report of the Parole Board 1985*, p. 8. The total number of prisoners serving determinate sentences who were granted parole in 1985 was 14,406, representing 62.3% of all cases considered.

[3] The New York State Division of Parole, *1982/83 Annual Report*, Albany, 1984, includes a short history of parole at pp. 2–15.

[4] P. J. Fitzgerald, *Criminal Law and Punishment*, Clarendon Press, Oxford, 1982, pp. 247–8.

From the inception of life imprisonment (as an alternative to transportation), convicted prisoners could be conditionally released on licence at the discretion of the Home Secretary, and this power was incorporated into the new arrangements contained in the 1967 Act.

In 1967, for the first time Parliament authorized the Home Secretary to release selected prisoners serving determinate sentences for imprisonment (i.e. for periods of time fixed by the courts) before the stage, normally after two-thirds of the sentence had been served unless remission had been lost, when they would otherwise have been freed. A minimum qualifying period of one-third of the sentence or twelve months (later reduced to six months), whichever expired the later, was set and an independent board, initially described as the Prison Licensing Board, was inaugurated to review the cases of eligible prisoners and to recommend to the Home Secretary whether or not, in the opinion of the Board, they were suitable for early release. Life-sentence prisoners or young offenders convicted of grave crimes and detained under Section 53 of the Children and Young Persons Act 1933 could be released at whatever time the Home Secretary thought fit, subject to a favourable recommendation by the Board, and after consultation with the Lord Chief Justice and the trial judge if available.

As originally introduced, the Bill envisaged parole without a Parole Board: namely, that the power to release would be vested solely in the Home Secretary. The notion of an independent board acting in an advisory capacity was a later refinement, added during the passage of the Bill through the House of Commons. The compromise that finally emerged on powers was that, whereas the Home Secretary would be unable to release a prisoner on licence without first receiving a favourable recommendation to do so, he would not be bound to accept such recommendations. When the Bill reached the House of Lords, an amendment moved by a former Home Secretary, Lord Brooke of Cumnor, supported by the Lord Chief Justice, Lord Parker of Waddington, proposed that the name of the new board be changed from Prison Licensing Board to Parole Board.[5] This was

---

[5] *Parl. Debates*, HL, 284 (5th ser.), cols. 435–8, 3 July 1967. A similar amendment had been moved at the Committee stage but was withdrawn.

partly to distinguish it from the authorities in the prisons (who were to have a say, but not a decisive role, in the selection process), and partly because of the sheer inappropriateness of the original title. In accepting the amendment at Report Stage, the Home Office Minister, Lord Stonham, admitted that the Board would be neither licensing prisons nor, strictly speaking, licensing prisoners, since the licence would be issued by the Home Secretary in those cases where he accepted the recommendation of the Board.[6] Having thus been present at the christening as well as the legislative birth of the new body, the House of Lords has always taken a close interest in parole and has provided three out of the four chairmen of the Parole Board so far.[7]

From the start, it was clear that the last word was to rest with the Home Secretary. It was to be his discretion whether or not to release any individual prisoner earlier than would otherwise be the case, and Parliament did not, either then or subsequently, transfer that power of decision out of the hands of the accountable Minister. It did, however, attach to ministerial discretion a new and independent process. If the Secretary of State agreed with the Parole Board's recommendation, the prisoner would be released on licence; if he disagreed, then the final decision was his as the Minister responsible to Parliament, rather than that of an autonomous board, following the general practice in the United States, Canada, and most other English-speaking countries. Ideals of fairness and consistency are not easily reconciled with the exercise of discretion, and there has been much debate about the extent to which the Parole Board's statutory responsibilities qualify or dilute the Home Secretary's discretion. In 1984 the issue was tested in the courts, and in *Findlay's* case the House of Lords held that the final decision, both on the period of time spent in custody and on the risk to the public should the prisoner be released on licence, rests with the Home Secretary. The law lords also ruled that, as the Minister responsible to Parliament and the electorate, the Home Secretary is entitled to take a broad view when exercising his discretionary

---

[6] *Parl. Debates*, HL, 284 (5th ser.), cols. 438–9.
[7] Lord Hunt was the founder chairman of the Board (1967–74), followed by Sir Louis Petch (1974–9), Lord Harris of Greenwich (1979–82), and the present author (1982–).

powers under the Act so as to have regard to the public interest and the public acceptability of his policies.[8]

Once released on parole, each offender is subject to a licence, issued to him personally, requiring him to report without delay to a specified probation office and to place himself under the supervision of whichever probation officer may be nominated for the purpose. The licence prescribes that his place of residence must be approved by the probation officer, who also needs to be informed of any changes in address or employment. It is a standard clause that the prisoner shall be of good behaviour and lead an industrious life, although the latter injunction is seldom interpreted literally. Nevertheless, the contribution made by the probation service in providing the all-important elements of compulsory supervision and emotional and practical support in the community should never be underestimated. It is the supervising probation officer who keeps his (or more often her) eyes open for warning signs, if necessary recommending recall to custody if a parolee breaches the conditions of the licence—often an indication that further offending may be imminent. In this way, although once again at liberty, an offender on parole is still serving the sentence of the court. The degree of freedom enjoyed is conditional rather than absolute, and the liability to recall is some deterrent against re-offending during the currency of the licence.

The historical origins of parole are worth rehearsing because of the misapprehensions that have grown up. The possibility of regaining the liberty that was forfeited by the commission of a criminal offence, punished by a sentence of imprisonment, before the expiry of the period of time set by the court means that strong feelings are evoked. Prisoners tend to see parole in black and white terms. To them, what matters above all else is whether they are inside or outside prison: the conditions attached to the licence seem insignificant in comparison. In their minds, too, the grant of parole is closely associated with good institutional behaviour and favourable reports from prison officers. Although this is an important factor, it has to be balanced with the gravity of the crime that attracted the sentence, the home circumstances, and the risk of re-offending, some of

---

[8] *In re Findlay* [1984] 3 WLR 1159–73.

these factors being outside the prisoner's control. Public attitudes also come into play. Neither in the United States nor (later) in Britain has parole ever been popular at the grass-roots level. There is little evidence to suggest that this situation is changing; indeed, most indications point the other way. Suspicion of a soft option lies at the root of popular attitudes to parole, while scepticism as to the fairness of the selection process leads many informed and otherwise well disposed people towards a more censorious stance.

Criticism of parole may be hostile and frontal (as when a further serious offence is committed by a prisoner on licence), or more subtle. Apparently small changes in emphasis can set the scene in ways that suit a particular outlook. I offer two quotations, neither of great significance in themselves, as examples of the thought transference that can occur from what is to what might be: 'The power of the Home Secretary to order the continued detention of any prisoner until the normal completion of his sentence . . .';[9] and, again, 'while it is certainly the Home Secretary's right to deny a lifer release on any grounds he chooses, . . .'.[10] Although the consequences to the prisoner are the same, it is an inversion of a positive power to release, conferred by statute on a Minister, to speak of the exercise of a negative right to order 'continued detention' or to 'deny' a prisoner release. References to the Home Secretary exercising a 'veto' when rejecting recommendations of prisoners as suitable for release are equally inappropriate.[11]

II

Each case of a parole-eligible prisoner which comes to the Parole Board for review is considered individually on the basis of full written reports. Voluminous dossiers are compiled from police information describing the nature of the offence and listing any previous convictions, reports from prison staffs and doctors with

[9] S. McCabe, 'The Powers and Purposes of the Parole Board', *Criminal Law Review* (1985), 489.

[10] M. Maguire, F. Pinter, and C. Collis, 'Dangerousness and the Tariff', *British Journal of Criminology*, vol. 24 (1984), 267.

[11] M. Wasik and K. Pease, 'The Parole Veto and Party Politics', *Criminal Law Review* (1986), 379–82.

first-hand knowledge of the prisoner, and reports from the probation service. The probation reports include any social enquiry reports that may have been prepared for the court and a home circumstances report, usually based on a personal visit to the address to which a prisoner proposes to go on release. There will also be a report from a member of the local review committee (LRC) who has interviewed the prisoner, as well as any written representations which the prisoner wishes to be considered by the LRC and the Parole Board. The Board is notified of the recommendation for or against release made by the LRC and is free to agree or disagree with either conclusion.

The Board is now so large a body, averaging over fifty members serving terms of not more than three years, that the size of the membership and the volume of cases preclude the possibility of the entire Board sitting together. Instead, panels of four members meet virtually every weekday in London, with regular (but fewer) meetings in Manchester and Cardiff. In 1985 almost 8,500 cases were reviewed—an increase of about 1,000 compared with 1984—at a total of 265 panel meetings.

A majority of Parole Board members have some professional qualification or experience of dealing with offenders. The Board is not, nor was it intended to be when set up, a cross-section of the general public. Out of a membership of fifty-five on 1 January 1986, there were seven judges (three High Court and four circuit judges), eight members of the probation service (seven Chief Probation Officers and one Deputy Chief Probation Officer), one retired Chief Probation Officer, and nine consultant psychiatrists. These three professional groups—judicial, probation, and medical—generally contribute about half of the total membership, with the remainder made up of independents including academic criminologists, former police and prison officers, educationalists, magistrates (one being the current chairman of Council of the Magistrates Association), barristers and solicitors with criminal practices, former members of local review committees and Boards of Visitors at the prisons, and voluntary workers with offenders. Members come from regions throughout England and Wales, and the main ethnic minorities are represented. On 1 January 1986, thirteen out of the Parole Board membership of fifty-five were women (24%). All members work part-time.

Appointments, normally for three years, are made by the Home Secretary—in the case of the chairman, with the approval of the Prime Minister. The judges are nominated by the Lord Chancellor and serve for two years, as do the probation members because of their heavy professional commitments. The chairman is paid a salary, and other members (with the exception of the judges) are paid a fee for attendance at meetings and for a day's reading of the documents in advance. The senior serving judge acts as vice-chairman. Although the annual work load has risen from some 6,500 cases in 1981 to almost 8,500 in 1985, the size of the Board has increased only from fifty-two to fifty-five over the same period.[12]

Despite these diverse backgrounds, a common outlook has evolved which enables members to reach agreement on their recommendations without undue division. Although panels of four can split equally, with two members in favour of parole and two against, this happens comparatively rarely. When a panel is evenly divided, the case will be deferred for further consideration by a fresh panel, to which a fifth member will be added to ensure that a decision is made. Each panel will consider twenty-eight or more cases of individual prisoners, some of which may have been reviewed previously. Normally, twelve months will elapse between reviews, although the Board may recommend an earlier review if it seems justified. With the exception of panel meetings when a smaller number of cases of life-sentence prisoners are considered, there is no indication of departmental or ministerial views. Home Office advisers sit in on lifer panels only. In addition to the main work, there is a daily flow of urgent business: licence revocations and recalls to custody, prisoners' representations against recall, and applications to vary licence conditions or to travel abroad. In certain cases, further information may have been requested from the probation service in order that a release plan may be referred back for specific approval by the Board.

Cases are considered in the light of six 'Criteria for selection

---

[12] There were 52 members of the Parole Board on 1 Jan. 1981 and 55 on 1 Jan. 1986. In the course of each year numbers fluctuate as new members join and old ones leave, not always at the same time. Thus the totals can be higher or lower on any given date.

for parole', which are published annually,[13] and the relevant criteria are noted by letters A to F, for the record, against any prisoner found unsuitable for parole. Reasons are not given to the prisoners or their representatives, and there is no provision for appeal. Prisoners may, however, petition the Home Secretary for another review. There is one exception to the policy of not giving reasons. Where a prisoner is recalled to custody after revocation of his licence, he is entitled to be informed of the reasons for his recall and to make representations in writing to the Parole Board. If the Board accepts his representations, he must then be re-released.[14] The power was used on fifteen occasions in 1985.

This brief outline of the composition and working practices of the Parole Board is entirely functional and does not answer the most conspicuous objections, which are that the prisoner is not seen in person, nor is he given the reasons why the Board considers him to be suitable or unsuitable for early release from custody. Criticism on these lines is both important and healthy; it deserves to be taken seriously and is discussed at some length later in this chapter. In the Home Office there is a litany (inspired more by the need for justification than by a desire for intercession) which describes parole as a system designed to permit the release of as many prisoners as possible, at as little risk to the public as possible, by means that are as fair and defensible as possible, at as little cost to the public as possible. It is not a bad definition, but it is one that calls for some interpretation.

III

Collective activity on such a scale generates its own momentum and provides its own satisfactions. The conscientiousness and application brought to bear is remarkable, and the inter-disciplinary nature of the Parole Board's membership leads to a constant broadening of outlook for those whose experience would otherwise be more limited. There is never any shortage of

[13] *Report of the Parole Board 1985*, pp. 15–18.
[14] The statutory provisions relating to the revocation of licences and recall to prison are contained in Section 62 of the Criminal Justice Act 1967.

well qualified people when the Home Secretary makes appointments to the Board, and many judicial members especially have regarded a term on the Parole Board as a unique opportunity to learn about what happens to offenders after they have left the courts, and the ways in which various forms of custodial treatment and release back into the community may relate to future offending. Valuable as this is, nevertheless, the parole system exists not to develop the virtuosity of those who are privileged to serve on the Board, but for the betterment of the many thousands of prisoners whose cases are considered each year.

In 1985 local review committees sitting at the prisons examined 22,912 cases.[15] Not all of these came to the Board, since, under powers contained in Section 35 of the Criminal Justice Act 1972 and Section 33 of the Criminal Justice Act 1982, parole for substantial numbers of shorter-sentenced prisoners is decided by the Home Office on the basis of a local recommendation only. The overall picture is that in 1985 a total of 14,406 prisoners serving determinate sentences of imprisonment were released on licence, 803 were recalled to custody for breaches of licence conditions, and about 6,000 were under supervision in the community at any one time. The average length of the licence was five months and one day.

Table 15 depicts a statistical profile of the growth of the parole system in England and Wales since its introduction in 1968. The first three columns show the rate of increase with the surges that occurred as a result of the extensions brought about by the statutory changes of 1972 and 1982 (the first releases under the 1982 Act taking effect from 1 July 1984). Another milestone was passed in 1975 when Roy Jenkins, Home Secretary for the second time, decided to use his discretionary powers to release more prisoners on parole, provided that this was not at the cost of a greater risk to the safety of the community. The method adopted was to distinguish between serious offenders, whose

---

[15] Local review committees are appointed by the Home Secretary to serve at each prison and youth custody centre. Their function is to review the cases of eligible prisoners at the establishment and make recommendations as to their suitability for parole. Under subordinate legislation, membership of an LRC consists of the governor of the prison (or his deputy), and not less than four other persons including a probation officer, a member of the Board of Visitors of the prison, and two independent members who are otherwise unconnected with the prison.

TABLE 15. *Release on licence, 1968–85*

	CJ Act 1967 S. 60	CJ Act 1972 S. 35	CJ Act 1982 S. 33	Prisoners released (determinate sentences)		Recalled to custody
				No.	% of cases considered	
1968[a]	1,157	—	—	1,157	11.8	30
1969	1,835	—	—	1,835	27.0	90
1970	2,210	—	—	2,210	28.3	138
1971	2,956	—	—	2,956	30.8	227
1972	2,915	—	—	2,915	32.7	237
1973	2,515	813	—	3,328	33.8	252
1974	2,826	676	—	3,502	35.4	270
1975[b]	3,106	923	—	4,029	42.6	311
1976	2,876	2,115	—	4,991	49.5	394
1977	3,192	2,018	—	5,210	50.4	538
1978	3,186	1,622	—	4,808	47.2	440
1979	2,833	1,925	—	4,758	46.8	421
1980	3,079	1,998	—	5,077	50.4	445
1981	3,355	1,916	—	5,271	54.8	544
1982	3,138	2,042	—	5,180	56.3	593
1983	3,443	1,903	—	5,346	56.1	512
1984[c]	3,330	2,767	5,789	11,886	62.3	512
1985	3,137	2,732	8,537	14,406	62.9	803

[a] Figures for 1968 are on the basis of recommendations made by the Board, as no count was made of the number actually released on licence. No mention is made in the 1968 Annual Report of the Secretary of State having rejected any recommendations.
[b] A policy statement about the extension of parole was made by the Home Secretary on 4 August 1975.
[c] 1984 figures reflect only six months of S. 33 releases.
Source: Annual Reports of the Parole Board 1968–85.

Release on Licence

further offending was likely to pose a grave threat to the public, and offenders who had committed less serious offences. It was his intention that more of the second category should be considered for parole after being reviewed by a local review committee, and without reference to the Parole Board as authorized by Section 35 of the Criminal Justice Act 1972 and agreed by the Board. The statistics from 1976 onwards reflect this change in policy. The large increases in 1984 and 1985 result from the reduction of the minimum qualifying period to six months. Section 33 cases, as they are known, include those prisoners serving shorter determinate sentences who are released under the powers contained in Section 35 of the Criminal Justice (CJ) Act 1972.

Prisoners serving sentences of life imprisonment have their cases reviewed in a way that is basically similar, except that the reporting is fuller and all cases are considered by a panel of the Parole Board including a High Court judge and a consultant psychiatrist with knowledge of treating offenders. There are, however, some procedural differences, of which the most important is that, since there can be no one-third point in an indeterminate sentence at which a prisoner may become entitled to a review, the stage when a lifer's case will be referred to the Board is decided by the Home Secretary. In doing so, he will take account of the nature and gravity of the offence, as well as the views of the judiciary about the period of time required to be spent in prison to mark the requirements of retribution and general deterrence. Both the original trial judge, if he is still available, and the Lord Chief Justice of the day must be consulted (there is a statutory obligation to do so), and Home Office Ministers will give weight to advice from this source in deciding when to refer the case of a lifer to an LRC for preliminary review. The first LRC review will normally take place about three years before the expiry of the period thought necessary to meet the requirements of retribution and deterrence, in order to give time to prepare a prisoner for release, for example by transfer to lower-security conditions, and in certain cases by a period of some months on a pre-release employment scheme living at a hostel. These moves will be sanctioned only after the degree of risk to the public has been assessed by the Parole Board.

Once this has been done, and where the Parole Board comes down in favour of setting a date for release, the final decision, on both the substance and the timing, rests with the Home Secretary. It is he who is the Minister accountable to Parliament, and it is upon his discretion that the authority to release depends. The release of a lifer, or indeed of any other prisoner before the end of his sentence (remission apart), is essentially an administrative, and hence ultimately political, act rather than a judicial one. The distinction is reflected in the practices adopted and lies at the heart of the institution of parole.

IV

In November 1983, Leon Brittan, then the Home Secretary, informed the House of Commons that, in view of public concern about violent crime, he intended to use his discretion in future in such a way as to ensure that murderers of police or prison officers, terrorist murderers, sexual or sadistic murderers of children, and murderers by firearms in the course of robbery would normally serve at least twenty years in custody.[16] It was a classic instance of a political response to a tide of opinion that had been articulated strongly in the general election campaign that summer. The new policy coalesced with and reinforced a hardening of attitudes which was already becoming apparent throughout the review process, resulting in fewer cases of life-sentence prisoners being referred to the Parole Board and fewer of those who were referred being recommended for release. The totals over a five-year period 1981–5 are shown in Table 16. The reasons for the steady decline in the rate of release (which, it should be remembered, has occurred while the overall population of life sentence prisoners has been increasing) are hard to identify. There have been no ministerial directions to the Parole Board or LRCs; members of both use their individual judgment in assessing each case on what they see as its merits. A large

---

[16] The text of the Home Secretary's statement was published in full in the *Report of the Parole Board 1983* (HC, 463), HMSO, London, 1984, p. 9, and in subsequent years. It had been foreshadowed in a speech at the Conservative Party conference at Blackpool the previous month.

TABLE 16. *Release of life-sentence prisoners, 1981–5*

	1981	1982	1983	1984	1985
Cases referred to the Parole Board	325	310	264	235	276
Cases recommended for release	142	133	110	93	76
Cases subsequently given release dates by the Home Office	138	128	95	84	63
% of life-sentence prisoners recommended for release	43.7	42.9	41.7	39.6	27.6
% of life-sentence prisoners given release dates	42.5	41.3	35.9	35.8	22.8

Source: Annual Reports of the Parole Board 1981–5.

majority of lifer cases are of homicide (murder and man-slaughter), but exclude most of the diminished responsibility cases discussed in Chapter 4. Some cases of arson attract life imprisonment each year, while the courts seem to be resorting more readily to the life sentence in incidents of particularly serious rape or other sex crimes, especially where the offender has a record of previous convictions for similar crimes and has not been deterred from re-offending by determinate sentences of imprisonment.

When reducing a life sentence imposed on a football hooligan to three years' imprisonment in 1986, the Court of Appeal gave guidance to judges in the Crown Court on the proper use of the life sentence. Other than crimes for which life imprisonment was mandatory, the court laid down that it should be reserved for exceptional cases of which the most frequent examples were where the crime was of particular gravity, or the defendant was likely to be a particular danger to the life or limb of others when at large, or where he suffered from some marked mental instability which might be cured by the passage of time but where no one could forecast how long that might be.[17]

[17] *R.* v. *Whitton,* Law Report in *The Times,* 20 May 1986.

Besides the changes affecting the release of life-sentence prisoners, the Home Secretary's 1983 statement heralded a more restrictive policy towards paroling convicted offenders serving over five years' imprisonment for offences of violence to the person or trafficking in controlled drugs. Where these circumstances applied, Leon Brittan said that he intended to use his discretion so as not to grant early release on licence except in circumstances that were genuinely exceptional, or where supervision for a few months at the end of the sentence was likely to reduce the long-term risk to the public. After consultation with the Parole Board, it was agreed that all of the classes of cases that had previously been scrutinized (i.e. excluding those dealt with under Section 35 of the Criminal Justice Act 1972) would continue to come to the Board after the initial review by the LRC. This remains the practice today, and cases of prisoners covered by the policy are customarily reviewed by panels of the Board with no less care than others, while special consideration is given to see if any genuinely exceptional circumstances can be identified, or whether grounds exist for recommending a few months under supervision at the end of a long sentence as an added measure of protection for the public.

The number of prisoners released from sentences in this category in 1984 and 1985 is shown in Table 17. A large majority fell under the second of the two provisos: namely, a short period on parole at the end of their sentences as part of the policy of risk reduction. Although it proved impossible to define 'genuinely exceptional circumstances' under the first heading, the Parole Board put forward forty-three such cases in 1984 and thirty-three in 1985. Of these, the Home Secretary accepted the Board's recommendation in twenty-three cases in 1984 and thirteen in 1985.[18] A higher figure was to be expected in 1984, the first year of the new policy, since the 'genuinely exceptional' provision was applied to associates found guilty of the same crime who deserved to be treated consistently with co-defendants who had already been released on licence, or recommended for release, before the change came into effect.

The policy changes affecting life-sentence prisoners, as well as those serving longer determinate sentences for crimes of

[18] *Report of the Parole Board 1984*, p. 4, and *Report of the Parole Board 1985*, p. 5.

TABLE 17. *Release on licence, by sentence length, 1984–5*

	Sentences of 2 yrs' imprisonment and over (all offences)[a]		More than 5 yrs' imprisonment (offences involving violence and drugs)	
	1984	1985	1984	1985
No. eligible for consideration[b]	11,496	11,946	1,241	1,520
No. recommended for release	6,120	5,894	204	278
No. actually released	6,097	5,869	189	258
No. released as a % of those considered	53.2	49.3	16.4	18.3

[a] The totals shown for sentences of 2 years' imprisonment and over (all offences) includes those sentenced to more than 5 years' imprisonment.
[b] Does not include prisoners who declined to have their cases considered.
Source: Annual Reports of the Parole Board 1984–85.

violence and drug trafficking, were challenged politically and in the courts, on the argument that not all prisoners were being given an equal opportunity to be considered for parole and that the Home Secretary was fettering the use of his own discretion. On appeal to the House of Lords, it was held that the Home Secretary had not stepped outside the territory within which a Minister can act in accordance with the powers conferred upon him by Parliament. *Findlay*'s case, to which reference has already been made, was evidence of the relevance of the doctrine of judicial review in testing and controlling the use of political power. On an application for judicial review of a ministerial decision or policy, the court's function is limited. It is not required to say whether it agrees or disagrees with what has been done; all it can do is to say whether it was in accordance with the law.[19] The House of Lords judgment, delivered by Lord Scarman, went well beyond the particular situation of the four applicants for relief (sentenced prisoners liable to spend a longer time in custody as a result of the new policy) whose plight had provoked

[19] *Report of the Parole Board 1984*, p. 5.

sympathetic comment. By such reinterpretation of the uncomfortable equilibrium between liberty and authority is the rule of law maintained.

The initial compromise between the powers of the Home Secretary and the role of the Parole Board has left an ambiguity which is thrown into relief when an application is made for judicial review. The formula arrived at in the Criminal Justice Act 1967 was that, while the Home Secretary was not able to release a prisoner without a favourable decision of the Parole Board (or, later, of the LRC in cases of shorter-sentenced prisoners), he was not obliged to accept such a recommendation. This must have sounded neat and sensible at the time, as well as satisfying those MPs who wanted to see a brake imposed by an expert board on the inclinations of any over-liberal Home Secretary. The practical effect, however, has been to obscure the distinction between a recommendation to release and a decision not to do so. In each year the total number of determinate-sentence prisoners recommended by the Parole Board as suitable for release who are not granted parole by the Home Secretary is small. In 1985, out of 3,293 cases of determinate sentence prisoners in which the Parole Board recommended release, the Home Secretary was unable to accept the Board's recommendation in only 25 cases, 20 of which fell into the restricted category of longer-sentenced prisoners convicted of crimes of violence or drug trafficking.[20] In addition, Parole Board recommendations for the release of life-sentence prisoners were not acceptable to the Home Secretary in 14 cases during 1985.[21]

In these cases, relatively few in number, the Home Secretary exercises his right to have the last word. But in the much larger number of instances where prisoners are considered not to be suitable for release after review by the Parole Board, the refusal is final. The distinction between a recommendation and a refusal is thus one of substance, and not merely a matter of words. When reaching a favourable conclusion the Board is making a recommendation; whereas when it finds a prisoner not suitable for

---

[20] *Report of the Parole Board, 1985*, p. 8.
[21] Ibid., p. 10. In one further case the Home Secretary was unable to accept a recommendation for release some time ahead of a recalled life-sentence prisoner following the consideration of his representations by the Parole Board.

parole it is making a decision. While it is hard to see how the courts can intervene to quash what is no more than an expression of opinion (provided it was reached in good faith and without procedural irregularities); equally, it could be argued that a refusal is a decision taken by an administrative tribunal affecting the rights of an individual, and hence a proper subject for review. Although somewhat pedantic to explain, confusion can be avoided by being clear about the principles inherent in the statutory framework. This is not to suggest that the High Court should provide an avenue for the further review of adverse decisions which disappointed prisoners seek to question or overturn. The sheer pressure of business, quite apart from the constitutional proprieties, would rule out any such possibility. The parole decision, whoever it is taken by, is an administrative one and not in itself justiciable. But if there is any doubt that the powers granted by Parliament may have been exceeded, or if there is obscurity, or if it is alleged that procedures sanctioned by subordinate legislation have not been complied with, then the courts have a right to step in. As the growing number of parole cases reaching the High Court in recent years has proved, they have not been slow to do so.

V

In disinterring the penal objectives that lie behind parole, it is possible to identify three recognizable layers. First, there is the reform of the man or woman who has committed a criminal offence which has resulted in a prison sentence: to encourage a sense of remorse and contrition and to seek to develop or strengthen the motivation not to offend again. Nearly all offenders, with the exception of some professional criminals, stop intentional offending at some stage in their lives, usually as they get older and their way of life becomes more stable; and it should be a prime aim of penal policy to do everything possible to accelerate and support this process. Next, there is the rehabilitation of the offender in the community: his acceptance by family and friends, and his search for a place to live and for some form of work or a regular occupation. As discussed in the previous chapter, for the generation immediately before and after the Second World War the objective of rehabilitation gave shape and purpose to the management and orientation of the prisons and the probation service. Treatment and training, particularly of longer-term prisoners and juveniles, was designed to provide those offenders who responded to the regime with a better chance of leading more decent and industrious lives after their prison sentence than before they came in. After-care of prisoners, by both the probation service and voluntary organizations, and later the idea of parole, fitted comfortably with this approach. By 1965 it was orthodox to claim, as the Home Office did in a White Paper in that year, that:

Prisoners who do not of necessity have to be detained for the protection of the public are in some cases more likely to be made into decent citizens if, before completing the whole of their sentence, they are released under supervision with a liability to recall if they do not behave.[22]

Contrast this with the tone of a comment from the Annual Report of the Parole Board for 1981:

Since the inauguration of the parole system in 1968, at the close of a decade in which great importance was attached by penologists to the principle of rehabilitation, those involved in penal affairs have become much more sceptical about the rehabilitative claims of the penal process, especially imprisonment. The parole system also began to reflect this change of emphasis, with decisions being based much more on a general assessment of the risk of future offending than on a narrower consideration of the rehabilitative effects of the prison experience.[23]

The sixteen years separating these two observations saw a noticeable shift away from the prospects for the rehabilitation of the offender towards an assessment of the risk to the public of further offending. This trend has continued over the past five years, although, as in archaeological sites, the remains of previous cultures, the reformative and the rehabilitative, are still evident, always below and sometimes showing above the surface. Penology cannot afford to jettison the values of past cultures. Instead, we should draw strength and inspiration from them, reinterpreting objectives that are still valid in the light of experience. Reform and rehabilitation are enduringly worthwhile aims, sometimes capable of realization if expectations are

[22] *The Adult Offender* (Cmnd. 2852), HMSO, London, 1965, p. 3
[23] *Report of the Parole Board 1981* (HC, 388), HMSO, London, 1982, p. 1.

not pitched too high, and always worth striving for. Humanity demands nothing less.

Today's objectives for parole are formulated more prosaically. All prisoners serving determinate sentences have to be released at some stage, normally after two-thirds of their sentence. When considering individual cases, the Parole Board and the LRCs are seeking to minimize the risk of re-offending, both during and after the parole period, by scrutiny and assessment of the nature and circumstances of the offence, the pattern of previous convictions, prison performance, and the proposed release plan. The prisoner's motives and attitude may still have a place, but these can only be subjective judgments.

The fundamental question confronting the Board and the LRCs is not simply whether a prisoner, if granted early release from custody, is likely to complete the licence period successfully without re-offending. Rather, it is whether in the longer-term interests of the public the risk that a prisoner will re-offend at any time in the future can be reduced or deferred by a period on parole. Is it wiser, in the particular circumstances of each case, to release a prisoner again into the community under compulsory supervision, subject to conditions relating to accommodation and behaviour, and liable to be recalled to prison in the event of misbehaviour or further offending; or to retain him in custody, typically for a few weeks or months longer, and then release him when his sentence expires without the element of supervision and support by the probation service which release on licence would have made available?

In roughly half of the more serious cases (those that have attracted sentences of two years' or more imprisonment) which it considers each year, the Parole Board decides that the public interest would be better served by a period on licence than by retention in custody until the expiry of the sentence. The more serious the crime, the less likely it is that parole will be recommended at first or second review. Although the available research is not conclusive, there are reasonable indications that a period of supervision on licence can be of positive benefit to offenders in assisting them to refrain from, or at least defer, further criminal activity. Bearing in mind that as many as one-third of all prisoners have nowhere to go on release, the provision of hostel or other accommodation found by the probation service can only help to loosen the ominous connection between homelessness and recidivism.[24] Studies of the comparative reconviction rates of prisoners released on parole with those serving similar sentences who are not paroled provides evidence to support the proposition that a period on licence helps to reduce the incidence and gravity of further offending.[25]

If this is the penological principle on which the contemporary practice of parole depends, what are the flaws that threaten, and could destroy, the organism? Three are especially corrosive: conspicuous failures on parole; persistent criticism of the processes for selection and recall; and judicial dissatisfaction with the procedures for the release of shorter-sentenced prisoners resulting from the reduction of the minimum qualifying period to six months.

VI

It does not require any very penetrating study of the workings of the penal system to be struck by the extent to which public policy has been influenced by events more than by dogma or the fruits of dispassionate examination. Crime and punishment is a notoriously sensitive area of policy and administration, and one that is at times unduly susceptible to public opinion, or to what is regarded by politicians as public opinion. Thus, unexpected incidents have attracted political responses. An example was the spate of prison escapes in the mid-1960s, which led to the Mountbatten inquiry into prison security in 1966, and later to the dispersal system. In setting up the Butler Committee on Mentally Abnormal Offenders, whose far-reaching report was noted in Chapter 4, the Home Secretary of the day, Reginald Maudling, was particularly conscious of the case of Graham Young, a poisoner, who in 1972 received four sentences of life imprisonment having being convicted on two charges of murder and two of attempted murder following his discharge from a

---

[24] See M. N. P. Ramsay, *'Housing for the Homeless Ex-offender: Key Findings from a Literature Review'*, (Home Office Research and Planning Unit Paper 13), HMSO, London, 1986.
[25] See *Review of Parole in England and Wales*, Home Office, London, 1981, pp. 9–11.

special hospital. Young had been sent to Broadmoor, subject to a restriction order, after being found guilty of three previous crimes of administering poison. In his memoirs, Maudling acknowledged that the further deaths resulted from a decision for which he took responsibility, and remarked that the burden on the Home Secretary in cases of this sort seemed to him a heavier one to carry than that of confirming a death sentence after due trial before a court of law.[26]

It is not usually the annual statistics of failures on parole, the number of licences revoked, and offenders recalled to custody that catch the public eye, but the exceptional and well publicized incident. By this yardstick, 1985 was a bad year in terms of parole failures. Two very serious crimes of rape, one followed by the murder of the victim, committed by men released on licence from sentences of imprisonment for previous offences of rape, provoked hostile comment in the press, from the Bench, and in Parliament. In the first case, the man had threatened that he would kill his victim if she went to the police when he had raped her the first time. After being released on parole, although closely supervised and attending a probation day centre, he returned, raped her again, and carried out this threat to murder her, setting fire to her flat before he left. In the second case, another man was no longer on parole when he raped and abducted a thirteen-year-old girl, although he had been released on three months' licence, initially with a hostel condition, earlier in the year. At the time of his further conviction, much publicity was given to the fact that he had served less than half of a four-year sentence resulting from previous rapes of girls aged fourteen and fifteen. Both offenders were sentenced to life imprisonment.

Rather different was the case of one of three men charged with the £26 million Brinks-Mat robbery at Heathrow Airport (mentioned in Chapter 3 as an example of large-scale organized crime), who was on parole from a twelve-year sentence when arrested and charged with conspiracy to rob. In the event he was acquitted, although his two co-defendants were convicted and each sentenced to twenty-five years' imprisonment. Both of them

---

[26] Reginald Maudling, *Memoirs,* Sidgwick & Jackson, London, 1978, pp. 162 and 173.

had stood trial with the parolee when he had previously been convicted, but on the earlier occasion it was they who were acquitted, and on the same grounds. There was thus a remarkable symmetry about the prosecutions and trials that followed each of these very serious robberies.

In the immediate aftermath of spectacular parole failures, it is futile to speak of the impossibility of eliminating completely the risk of re-offending, ruminating that the human mind is too mysterious and unpredictable for there to be any certainty, pointing instead to the reduction in risk that can result from a period of supervision in the community. Although valid, the reasoning is too subtle, at once apologetic and detached, to be an effective counter to the loud and angry voices that articulate public reactions of outrage and disgust. What is called for is the patient and insistent reiteration that, while parole cannot eliminate the risk of re-offending, it can help to reduce it, so affording a greater measure of protection to the public. It is beliefs of this sort that need to take root in the public consciousness.

It may be an over-simplification, but it is not a distortion of the truth to say that, whereas the general public is concerned with the prisoners who are granted parole, the penal reformers are more concerned with the prisoners who are refused parole—and why. The way selection is made has been questioned, especially by those who are not released on licence, from the start. On the face of it, there does seem to be a basic unfairness in procedures that do not enable the prisoner to be heard in person, or to be represented by a friend or legal adviser, when so much is at stake. Nor is it easy to justify the practice of not giving reasons for the decisions that are reached. On another tack, some criminologists query whether, for all the care that is taken, subjective assessment of individual prisoners on the basis of written reports produces any markedly different results in terms of re-offending than random selection. Critics come together in a dislike of decisions of such profound significance, both to the individual prisoner and to the wider community, being taken in private, by appointees of the Home Secretary, with officials preparing the ground, and without any provision for accountability or appeal. Moreover, it is argued that consistency cannot be achieved, and that chance plays too large a part. Is it defensible that, where there are two prisoners whose circumstances are broadly similar, one may be regarded as suitable by a panel of the Board on one day of the week, while the other may be rejected by a separate panel meeting the following day? Why is it, then, faced with such plausible objections, that the working practices are as they are?

VII

The explanation lies in the differing nature of judicial and advisory functions. The point was well put when the legislation to introduce a parole scheme was before Parliament. Roy Jenkins, the Minister responsible for the passage of the Criminal Justice Bill in 1966/7, warned against procedures that would be:

very similar to a second trial, carried out before a judicial chairman in accordance, no doubt, with carefully laid down rules, probably with the prisoner legally represented, perhaps with both sides legally represented. In my view, this would be a far too legal and far too formalistic procedure. I do not think that the decision as to whether a prisoner should be released on licence is in essence a justiciable point. The essence of the decision which has to be taken is whether, looking some distance into the future and making the best forecast which can be made, the prisoner is more likely, if released, to lead an honest life and to trouble the public less than if he is kept in prison for some further period. That is a very important decision to take, but it is not a justiciable decision. It would be quite wrong to try to have a highly formalistic legal proceeding to deal with it.[27]

The procedure which the Parole Board adopted when it was first set up, and which has continued virtually unchanged ever since, reflects this approach. The informal working methods are those of an expert advisory body rather than a judicial tribunal. The fact that there is no power to order the release of any prisoner, unlike a Mental Health Review Tribunal when considering the discharge of a restricted patient under the 1983 Mental Health Act, is crucial. What Board members are doing is to review the merits of each particular prisoner's individual situation and to recommend, in the light of the available information, and drawing on the experience of the reporting officers,

[27] *Parl. Debates*, (HC), SC (1967–8), 2, col. 736, 7 Mar. 1967.

if they regard him as being suitable for parole. In an over-whelming majority of cases, the Home Secretary agrees with the recommendation, but a recognition that the final decision on release lies elsewhere points inescapably towards an advisory process.

The prisoner has an opportunity to put his case to a member of the LRC who interviews him at the prison before the first stage of a parole review. Both the LRC and the Board will consider a report of that interview, together with any written representations which the prisoner may wish to submit. Were the prisoner to be present in person when his case is scrutinized, the implications would go deeper than is apparent at first sight. Apart from the formidable practicalities and the cost of some 23,000 prisoners appearing before the LRCs and/or the Parole Board each year, there are libertarian overtones. Not all prisoners are equally articulate; indeed, as the Home Office *Review of Parole in England and Wales* remarked in 1981, the prison population contains at the one extreme some of the most plausible people in the country and, at the other, some of the most inadequate and inarticulate. The review looked back at the experience of the boards whose function it was to decide when a preventive detention prisoner could be released. This indicated that the inarticulate and inadequate prisoner felt himself to be at a disadvantage when appearing before a board, and the thought that his release might depend on his showing could impose unacceptable strains upon him. On the other hand, the glib and articulate prisoner was thought by other prisoners to have an unfair advantage. It is only natural that the inarticulate man or woman, faced with the prospect of appearing in person, should look for help in the presentation of their case. Friends or legal representatives would be brought in, leading in turn to pressure for the lawyer to address the committee or board in order that his client's case is put to the best possible advantage. One lawyer breeds another, the Crown would seek legal advice on how to respond, and in no time at all the old, familiar, and much criticized adversarial system would be in full swing. As in court, the prisoner would find himself once again more of a spectator than a participant.

The giving of reasons would be an irrevocable step towards a judicial system. This issue has been before the courts, with the

Court of Appeal in 1981 holding that there was no duty on the part of the Parole Board or LRCs to give the reasons why they had declined to recommend a prisoner for release, and that consequently they had not acted unfairly towards a prisoner in refusing to inform him of their reasons.[28] The result of this case was to confirm what had been implicit before, namely, that, if the Board did decide to disclose the reasons that had led it to recommend in favour or against parole, then the reasons could be called into question in the courts. If the court was not satisfied with the reasons, the outcome of the review could be quashed.

In the same way as any other public body the Parole Board has an obligation, legal as well as moral, to comply with the rules of natural justice. A prisoner whose licence is revoked is entitled to be told of the reasons why he has been recalled to custody. Although this is a statutory entitlement, it is one that has led to difficulties when the reasons are challenged. Then the Parole Board is in a quandary. The probation service or the police may give one account of a parolee's behaviour while on licence; he may give quite another in his own representations against recall. Who to believe? There are no witnesses, no rules of evidence. An advisory committee is not equipped to establish the truth in such cases. In 1984 the Parole Board was in the Court of Appeal again on the application of a prisoner who had been released on licence and unsuccessfully sought disclosure of reports leading to his recall to custody. In this case the Court of Appeal held that the application of the rules of natural justice would vary according to the tribunal in question. In many cases a tribunal would be under a duty to disclose adverse material. But that did not apply to the Parole Board in exercising its advisory function. Although the Board had a duty to act fairly, it had no duty to disclose to the prisoner adverse material in its possession.[29]

Thus, the advisory status of the Parole Board determines the procedures it has adopted. To go down the road of having the prisoner presenting his own case or being legally represented, to give reasons that would be open to challenge, or to grant rights of appeal, reasonable as all this sounds, would be to transform the basis of the present parole system. No doubt a judicial or

[28] *Payne* v. *Lord Harris of Greenwich and another* [1981] 2 All ER 842 *et seq.*
[29] *R.* v. *Chairman of the Parole Board and Secretary of State for the Home Department ex parte Gunnell,* Law Report in *The Times,* 7 Nov. 1984.

quasi-judicial alternative could be found. There would be pros and cons to be hammered out. But it is not the system that Parliament has devised and approved. During the passage of the 1982 Criminal Justice Bill, amendments requiring reasons to be given were debated and defeated in both Houses.

Is this any more than heartless rigidity, it may be asked, putting the niceties of the law above the instinctive promptings of conscience and common sense? In the past, Parole Board members and prison administrators have acknowledged that the policy of not giving reasons can fuel a prisoner's sense of grievance when turned down for parole. It was argued that bitterness and cynicism about the fairness of the whole procedure could result, and that, if the genuine reasons were not communicated, prisoners would invent their own, the consequences of their speculation sometimes being more damaging than the truth. There was a recognition too of the disillusion and frustration that can spread, enveloping prison officers as well as the prisoners, and their families and dependants, when parole reviews are unsuccessful.

The prison officer who recommends a prisoner for early release on licence does so on the basis of his observation of the man and his experience of many other offenders. Thus he has a measure against which to assess individual cases, but one that can be of only limited value, since by definition prison must be an artificial setting. It remains a truism that good prisoners, often conditioned by previous sentences, do not necessarily make good citizens. The Parole Board has a responsibility for what happens outside the prison walls rather than for the maintenance of that delicate relationship between inmate and staff upon which prison order depends. One of the objections to the Borstal system of indeterminate custody for young offenders was that it related the release of the trainee too closely to his performance in the Borstal, at once giving the staff a power over him that could be misused and denying the community a broader assessment of the risk of his potential for re-offending.

During the mid-1970s, in response to the criticisms that had been expressed, an experiment was conducted. Since not all Parole Board or LRC members reached their decisions by the same route, a standardized list of factors was drawn up—sixteen in all, with four more for lifers—representing the principal 'causes for concern' which could lead to a recommendation against parole. Over a twelve-month period, LRC members at a sample of prisons were invited to make a choice from the list; but despite regular briefings, they found this as hard to do at the end of the trial period as at the beginning. In those cases that were further reviewed by the Parole Board, the Board was asked to select its own causes for concern without knowledge of those that had been chosen by the LRC. In about a quarter of the cases considered by both the Board and the LRC, different causes for concern were given. In an attempt to test the likely impact on the prisoners whose cases were being reviewed, prison governors were informed in confidence of the selected causes for concern and asked what the effect would be likely to have been, if disclosed, on the prisoner, his family, and his attitude towards the staff. Although necessarily subjective, and based on no more than five LRCs and five different categories of prison, the results were once again disappointing. In the words of the Home Office report, 'Overall the assessments suggested that while in the large majority of cases behaviour and attitude to staff would be unaffected, in about ten per cent of cases attitudes would improve and in ten per cent they would deteriorate.'[30] In 1979 the 'Reasons' experiment was abandoned, and the Parole Board, perhaps reluctantly, advised the Home Secretary that the giving of reasons was incompatible with the parole scheme as it stood.[31]

Although to some critics the present way of doing things represents an outmoded and arbitrary relic of a largely discredited treatment model, the case for a different and possibly more judicially conducted process must be argued in Parliament. As has been shown, there are legitimate considerations to the contrary and the present arrangements contain some useful safeguards for the rights of individual prisoners. Not the least of these is access to the courts by way of a prisoner's application for judicial review in appropriate situations. To gnaw away at the supports will risk causing the collapse of the whole edifice. We should always remember that, although the way in which

[30] *Review of Parole in England and Wales,* p. 19.

[31] A statement of the Parole Board's view on the giving of reasons for the refusal of parole was included as Appendix 5 of the *Report of the Parole Board 1979* (HC, 651), HMSO, London, 1980, pp. 42–3.

selection is made is an important aspect of parole, with implications for fairness and natural justice, it is still only the mechanical part of what is one of the more enlightened and functionally useful penal policies of recent times. Moreover, the action has moved away from the issue of selection, pursued with such intensity for so long, to a more specific and unexpected development.

VIII

It was paradoxical that the reactions provoked by the changes in parole policies made in the autumn of 1983 centred almost entirely on the restricted opportunity for parole for prisoners serving sentences of more than five years for offences of violence and drug trafficking, and on the way life-sentence prisoners would henceforth be reviewed for release. But there was a third policy decision announced at the same time: this was to implement the power contained in Section 33 of the Criminal Justice Act 1982 so as to reduce the minimum qualifying period for parole consideration from twelve months to six. This change was accepted without question, and indeed attracted very little comment at all. While it was noted that the number of prisoners eligible for parole could be expected to rise sharply as a result, there was no public or parliamentary response beyond an acknowledgement that an anomaly had been rectified in that the previous arrangements unfairly favoured the longer-sentenced prisoners, who had committed the more serious crimes, at the expense of the shorter-sentenced (and less serious) offenders, who were denied the chance of early release.

The change came into effect on 1 July 1984, when nearly 2,000 newly eligible prisoners were released on licence. This did not go unnoticed, and brought a welcome (if temporary) measure of relief to the grossly overcrowded local gaols. Simplified procedures were devised to take account of the less serious nature of the offences in the lower sentence range and of the foreshortened time scale. Prisoners serving sentences of less than two years' imprisonment were reviewed by the LRC at their prison against what the Home Office circular referred to as a 'presumption in favour of parole'. Those found suitable would not normally be referred to the Parole Board, but would be released by the Home Secretary under powers deriving from Section 35 of the Criminal Justice Act 1972.[32]

It should be said that both the Parole Board and the judiciary had supported this extension and the measure of redistribution it involved between the work of the LRCs and the Board. In its Annual Report for 1983 the Parole Board had forecast an increase in the number of prisoners qualifying for parole consideration each year from about 10,000 to up to 23,000 (including young offenders who became eligible under new sentencing provisions in the 1982 Act).[33] What was not forecast, however, was the widespread dissatisfaction to which the new procedures would give rise. The presumption in favour of parole led to an erroneous but widespread belief that releases under Section 33 amounted to 'virtually automatic' parole. Even the *Criminal Law Review* lent its authority to this description,[34] despite the fact that approximately one in every four prisoners considered was not recommended for release.[35] Within the LRCs, greatly expanded in membership to handle the new load, disillusion grew as the sketchy facts before the committees and the time available for their consideration came increasingly to be regarded as inadequate. The situation had not yet degenerated into a caricature featured in the *New York Times*,[36] but it was moving in that direction. In one American state, it was claimed, lack of information about an offender or his offence led to parole decisions being based on dialogue that went like this:

Board member: Hi there. What's your name?
Prisoner: Joe Smith.
Board member: How long you been here, Joe?
Prisoner: 'Bout three years.

---

[32] The only exceptions are where the recommendations of a particular LRC seem to be significantly out of line with those of LRCs at similar establishments. In these circumstances a batch of cases may be referred to the Parole Board on grounds of parity for further consideration and the LRC is informed of the outcome. *Report of the Parole Board 1984*, p. 3.

[33] *Report of the Parole Board 1983*, p. 1 (para. 3 as corrected).

[34] 'The Criminal Justice Act 1982: Before and After', *Criminal Law Review* (1985), pp. 413–14.

[35] See Letter to the Editor from the chairman of the Parole Board, ibid., pp. 687–8.

[36] *New York Times*, 4 Apr. 1986.

Board member: Are you staying out of trouble?
Prisoner: Yes, Sir.
Board member: Do you have a job lined up when you leave here, Joe?
Prisoner: Sure do.
Board member: What about a place to live?
Prisoner: Gonna live with my sister.
Board member: Think you can keep a clean slate on the outside?
Prisoner: Yes, Sir.
Board member: Okay, Joe. We don't want to see you in here again. Good luck.
Prisoner: Thank you, sir.

The careful scrutiny of individual cases on the basis of full reports, and the conscientious and thorough assessment of the risks and benefits involved in early release, which had been the hallmarks of the parole scheme for fifteen years, were being jeopardized by the more perfunctory approach adopted by LRCs, with the encouragement of the Home Office, in an attempt to streamline the system.

During 1985/6 the administrative procedures for the release of shorter-sentenced prisoners serving terms of less than two years' imprisonment were subject to review by the Home Office, with the co-operation of Parole Board members. Certain improvements were identified, particularly in the quality of the documentation made available to the LRCs to enable them to make a fuller assessment of the nature of the original offence. In 1986 the Parole Board reconsidered its criteria for selection for parole, which are published annually and circulated for the guidance of LRC members. The references to any presumption in favour of parole were deleted and instead the emphasis was placed, as in all other parole decisions, on seeking to reconcile benefits and risk. There seems no good reason why, in Section 33 cases as in others where the parole period will be longer, the possible benefits to the offender and to the community of some time spent on licence under the compulsory supervision of the probation service, and subject to recall in the event of breaching the licence conditions, should not be weighed against the possible risk to the public of releasing the offender earlier than would otherwise be the case.

These changes should help to meet some of the concerns that have been expressed, particularly by the many public-spirited men and women who serve on local review committees at over

100 prisons and youth custody centres throughout England and Wales. They should also contribute to better informed recommendations. But such marginal adjustments do not go far enough to reverse the tide of judicial disquiet emanating from the Crown Court where the Section 33 procedures for paroling prisoners sentenced to more than nine months' and up to two years' imprisonment apply to a majority of all prisoners given a custodial sentence for an indictable offence. Circuit judges and recorders, individually and at sentencing conferences, have insisted that parole for shorter-sentenced prisoners has become 'virtually automatic', and that LRCs were under official instruction to recommend release unless there were compelling reasons to the contrary. Some judges contended that parole has been granted too readily to certain types of offenders, such as young recidivist burglars, and that proper consideration was not being given to the nature of the offence or the risk to the public of re-offending. Most powerful of all was the argument that the differentials in judicially imposed sentences were being eroded, sometimes even nullified, by executive action. A noted Cambridge criminologist denounced the scheme as reducing the process of judicial sentencing to 'a complete farce',[37] and his language, although strong, found an echo among sentencers. This line of censure is always damaging for any parole system, and it can undermine years of patient work in building up judicial confidence.

The requirement that a prisoner should serve one-third of his sentence or six months (not including any time spent on remand), whichever is the greater, means that differentials are flattened out. If granted parole under the arrangements applying to the release of shorter-sentenced prisoners since July 1984, there is no difference in the actual amount of time spent in custody by prisoners sentenced to twelve months, fifteen months, or eighteen months; each will be released at the six-months stage,[38] although the length of the licence varies. Six months is also the period served by a prisoner with full remission whose nine-month sentence does not attract parole eligibility. So what is

[37] David Thomas, 'Parole and the Crown Court', *Justice of the Peace*, 149 (1985), 344. Dr Thomas repeated his criticisms for a wider audience in the *Guardian* on 16 May 1986.
[38] Assuming no loss of remission.

the point, it is asked, politely but persistently, of judges agonizing over whether to pass a sentence of nine, twelve, fifteen, or eighteen months if the chances in the end are that most prisoners in this range will in fact serve exactly the same amount of time in custody whatever the sentence of the court? The fact that the period in custody no longer reflects a comparative measure of culpability is of particular relevance when co-defendants (for example, a gang of youths) receive different sentence lengths in this range from a trial judge wishing to distinguish relative degrees of responsibility for a particular offence or offences. Further inequities result from the rule that time spent in prison on remand before conviction or sentence, although counting as part of the total sentence to be served, is disregarded for the purposes of the minimum period of six months required to be spent in custody before any release on parole can take place. Although not new, since differentials were also eroded when the minimum qualifying period was twelve months, the argument is emotive and one that caught the mood of many sentencers.

Some judges may conceivably have been motivated less by the imperfections resulting from a shortening of the minimum qualifying period than by a more deep-seated conviction that the sentence of the court should not be interfered with, other than in the most exceptional circumstances, by any non-judicial body. Only the sentencing court, it is argued, has heard all of the evidence, has seen the defendant in person, has considered his previous record, and decided on the punishment he should endure. Considerations of deterrence, both specific and general, may have been in the judge's mind or featured explicitly in his remarks when passing sentence. Appeal provisions exist to correct any errors or inconsistency in the type or length of sentence imposed. What possible justification can there be, certain judges and criminal lawyers enquire, for setting aside all this by allowing a mixed bag of well-meaning people, without first-hand knowledge of the offender or what he did, to conclude that the sentence of the court should be foreshortened? Understandable as such opinions may be, and strongly held as they undoubtedly are in some quarters, they are based on questionable grounds of constitutional principle.

Long before the parole scheme was ever thought of, there was tension between the courts and the executive. There is nothing new about it; indeed, a sense of rivalry and jealous independence is a mark of a society in which the citizen may anticipate getting fairer treatment at the hands of both than if either one had achieved unchallenged sway. In English legal history the Crown has preserved its prerogative to lighten the sentence of the court, whether by way of commuting a death sentence, or granting a ticket of leave from transportation, or releasing an offender from prison before remission is due. Remission, when put onto a statutory basis, represented the first curtailment in the duration of custodial sentences to be authorized by Parliament. Parole followed later, once again a discretionary power of a Secretary of State specifically sanctioned by Act of Parliament. The historical point to grasp is that 'the criminal courts have never, even in theory, been in a position to decree how long a prisoner *must* be detained, but only how long he *may* be detained'.[39]

Others besides judges may claim to speak in terms of the public interest. Sentencers cannot take account of the effect of a prison sentence, nor of an offender's state of mind or future prospects, sometimes many years after conviction. Punishment must have a place in any society, and the courts of justice are its proper instrument. But the imposition of punishment by the courts, with all the functional and symbolic overtones that accompany it, cannot eclipse a greater imperative still: that the public is entitled to the maximum degree of protection from criminal offending that can be devised on its behalf. Since this objective must include a reduction in the incidence and gravity of further offending by those who have previously been convicted (a regrettably substantial category), some system of post-sentence review to assess potential risks and benefits is required.

It is likely, however, that it was the practical consequences of the Section 33 change, rather than any doubts about the policy decision to extend parole eligibility to a larger number of shorter-sentenced prisoners, which caused the Lord Chief Justice to intervene. Eschewing quiet words in Whitehall corridors or whispers in ministerial ears, Lord Lane, as befits the head of an independent judiciary, made his views public in a forthright speech at the 1985 Annual Lord Mayor's Dinner to HM judges. The changes in the rules on parole, he said are:

[39] Professor Nigel Walker, unpublished memorandum for the Parole Board, 1986.

increasing the gap—already great enough—between what the court orders and what happens to the criminal in fact. Has not the time arrived when the very existence or place of parole in our system of detention and release of prisoners should be subjected to a strenuous review? Some think the time for that is long past.[40]

The sequence of events is not yet completed. The fuse lit in the Crown Court, which blazed into flame at the Mansion House, is still smouldering, and no one can yet predict the outcome. A wider review, maybe covering remission and policies towards remand prisoners (discussed in the previous chapter), as well as parole, would be one possible response. Others may argue that something more radical and immediate is called for. The issues are a great deal more complex than they appear at first sight, and the constraints imposed by prison overcrowding limit ministerial freedom of action.

I end with this thought. Parole has never been a popular device. The system is intricate, and the benefits to society have a low visibility. It is without friends beyond the prisoners—now many of them—who obtain early release, and a handful of people in the prisons, in the probation service, and in the community who work alongside them for the greater protection of the public. Although the voices of the critics are not in unison, the effect can be cumulative. There is little public sympathy for offenders, nor are many imaginative and determined efforts made to comprehend the causes of crime—how it is committed and how it can be best countered. Two of the three legs which uphold the institution of parole can hardly be regarded as secure. The general public is largely antagonistic, while penal reformers, backed by a significant strand of informed opinion, are disillusioned. The third leg is judicial opinion. In the United States, when the judges turned against parole in favour of reforms in sentencing practice, it sounded the death knell. We cannot afford to let that happen here.

---

[40] Transcript of a speech by the Lord Chief Justice at the Mansion House, 9 July 1985. Lord Lane has first-hand knowledge of the parole system, having served on the Parole Board 1970–2, and as its vice-chairman in 1972.

9

Prevention and Crime Reduction

I

For too long, crime prevention, like parole, was looked on as a soft policy—all very well in its way, but insufficiently stern and positive to be accorded a place in the front rank of the campaign against crime. Over the last few years the perception has been changing. In Chapter 6 we have seen how prevention is emerging as the most promising response to drug abuse, particularly by young people. Preventive strategies extend beyond education, exhortation, and moral example, aiming at more effective protection of property from burglary, theft, and vandalism. There are also gains to be obtained in terms of personal safety, as well as worth-while side-effects.

Out of all recorded criminal offences each year, an overwhelming proportion are against property. Home Office estimates put the figure as high as 95%.[1] Many of these are opportunistic crimes, in the sense that they were not deliberately planned in advance with a specific target in mind. Most offenders do not go out in the morning with an intention to burgle a house or steal a car; the opportunity to do so occurs in the course of the day. Some of the more purposeful among them, it is true to say, may be on the look-out for suitable targets (making opportunities, rather than simply taking advantage of them),[2] but either way, the offence is much easier to commit if cars are left unlocked and premises insecure. One survey of domestic burglaries indicated that in 30% of the cases no forcible entry was involved, and it is probable that the proportion of unlocked

[1] *Criminal Justice—A Working Paper* (rev. edn), Home Office, London, 1986, p. 8.
[2] M. Maguire, 'Burglary and Opportunity', Home Office, *Research Bulletin* 10, HMSO, London, 1980, pp. 6–9.

cars that are stolen is even higher.[3] So the scope for prevention, in terms of improved security, is patently very large. This is the main thrust of recent initiatives, some of the most significant of which are outlined in this chapter.

But prevention means more than owners taking greater care of their property and making it harder for law-breakers to steal it. City-centre violence, especially around pubs and clubs and associated with the excessive consumption of alcohol, violence on public transport, and football hooliganism are all instances where the personal safety of the citizen is put at risk. Preventive measures can reduce that risk. Furthermore, there is the important phenomenon of the fear of crime, already noted in the opening chapter, an anxiety which can disfigure and restrict the lives of people, especially the elderly and vulnerable who are living alone. The fact that they may be statistically less at risk than other groups, such as younger, single men roaming the city centres, does nothing to reassure those who are afflicted in this way. When a crime actually occurs, whether the householder is self-confident or fearful, the effects can be unpredictable. Unlawful intrusion into someone else's home is in every sense a violation, and the emotional harm, deepened if the victim has the misfortune to be present at the time of a burglary or if violence is used, can be real and lasting.

Looking behind the parties directly involved, the offender and the victim, crime prevention is a policy with much to recommend it as a path to a better life. If squalid and decaying housing estates can be modernized and converted into safer places to live, then more dwellings will be let, more rents paid, more tenants contented, and the physical environment improved. If new housing developments are designed in such a way that security is a factor from the start, then some of the mistakes of the past will be avoided. If greater provision is made for youth and community facilities in deprived areas, there may be less opportunity or incentive for vandalism. If those living in adjacent dwellings join together to form watch schemes, keeping an eye on each other's homes and having contact with the local police, they are not only helping to reduce crime in their own immediate

---

[3] *Criminal Justice—A Working Paper*, p. 8.

localities, but are fulfilling the timeless role of the good neighbour with all the social and personal satisfaction that can bring.

Policies aimed at preventing crime have a value going beyond the purely functional one of reducing the incidence of criminal offending. They represent a polarization of good and evil and a strengthening of the idea of community. This may explain why crime prevention as an objective of criminal policy has attracted relatively little controversy over the last few years, being embraced with equal fervour by all of the main political parties, as well as by statutory and voluntary agencies. Theoreticians may argue over displacement and rational choice,[4] and politicians over poverty, unemployment, and social divisions as underlying causes of offending; but these well-worn themes may be overtaken, for at the grass roots the first intimations of a new divisiveness are beginning to emerge. Too strident an emphasis on the perils of crime and the need to guard against them, it is claimed, can reinforce the fear of crime and generate a fortress mentality. Organized schemes may be diverted into becoming pressure groups for punitive sentencing, or may concentrate on protecting the property of the more privileged and prosperous to the exclusion of the personal safety of the less fortunate. Isolated, but plausible, anecdotes circulate that vigilantes are being employed to harass certain groups such as black people or youths standing about peaceably (but arguably suspiciously) on street corners.

This is the stuff of which political controversy is made; but the end (crime prevention) is such a great prize, and the means (communal action) so virtuous, that it would be regrettable if political conflict were to cloud one of the more promising vistas in the control of crime. The most important aspect of all is that practical action by individuals is an integral part of this response to crime, substituting for feelings of frustration and helplessness a determination to find ways to fulfil public aims that simultaneously and directly benefit the private citizen.

---

[4] See D. B. Cornish and R. V. G. Clarke, 'Situational Prevention, Displacement of Crime and Rational Choice Theory', and G. Trasler, 'Situational Crime Control and Rational Choice: A Critique', in K. Heal and G. Laycock (eds), *Situational Crime Prevention: From Theory Into Practice,* Home Office Research and Planning Unit, HMSO, London, 1986, pp. 1–24.

II

Progress has been driven by dual mainsprings, national and local. At the national level, priority has been given to the generation of a more crime-conscious awareness in the central departments of government, charging them with a responsibility to formulate and communicate crime reduction messages in their respective fields. This approach stemmed from a recognition that criminal offending was not something with a life of its own, apart from the community in which it took place, but was an aspect of life within it. An Interdepartmental Group on Crime, chaired by the Permanent Secretary at the Home Office in 1983, analysed the problem in these terms:

Crime is a complex phenomenon with many roots. There are contributory factors associated with law-breaking in the character and background of individuals; and there are other factors affecting the incidence of crime in the physical environment. The services concerned with the education, guidance, or support of individuals and families are already a positive force for good: but it is clear that they could do still more to promote responsible behaviour, positive attitudes and respect for the law. More could also be done to ensure that the environment is not conducive to anti-social behaviour and offers adequate opportunities for non-criminal activities. For effective action there is a need to look at the physical design and planning of the environment and the provision and management of local services, both public and private. There is an unexpectedly wide range of measures that can be taken to reduce crime, some of them simple and involving little additional cost.

The continuing objective must be to ensure that the opportunities that undoubtedly exist for effective action against crime are recognized and exploited by every Department and local service involved. This means that institutions must be organized to work effectively with one another and in ways that maximise their shared potential for controlling crime ... The main participants in this are local government services—housing, environmental services and planning, social services and education; the police and the probation services, and the voluntary sector.[5]

The report of this Group brought out the importance of the setting in which crime occurs, stressing that crimes result not only from the motivation of the offender but also from the situation in which he finds himself. Thus, if the situation can be

---

[5] *Crime Reduction,* Report of an Interdepartmental Group on Crime, Home Office, London, 1984, p. 1.

altered to his disadvantage, for example by improved physical security, better lighting and alarm systems, more caretakers or entry-phones to flats, and increased chances of detection, then the opportunities for crime will be reduced. This is what is meant by 'situational prevention', based on the premiss that some sorts of crime can be reduced through the management, design, or manipulation of the immediate environment in which it occurs.[6]

To implement these ideas and keep up the momentum, a specialist Crime Prevention Unit was established in the Home Office in 1983, and this lively unit has made much of the running since then. Its task has been to co-ordinate and disseminate information on crime prevention, bringing together research findings and practical experience, generally prodding and chasing those who have the capacity to make things happen in the private sector as well as in the public service. By 1986 the Unit had twenty-four full-time staff and was responsible for planning two high-level national seminars at 10 Downing Street, the first chaired by the Prime Minister in January and the second by the Home Secretary six months later. These seminars were a good example of how the presence and personal interest of the most senior of all Cabinet Ministers can keep an issue in the public eye, drawing a correspondingly heavyweight representation from a range of selected organizations to discuss and carry forward practical measures for preventing crime. As a by-product of the Downing Street seminars, a Ministerial Group on Crime Prevention was established to co-ordinate crime prevention policies and central government activity against crime. Further regional conferences and meetings were arranged during 1986 in Norfolk, Humberside, and Hampshire, backed up by discussions with Chambers of Commerce at national level.

Alongside the Ministerial Group is ranged the Home Office Standing Conference on Crime Prevention. This body, now chaired by a Home Office Minister, has been in existence for some years, although it has recently been strengthened and given a more prominent role. One of the most important earlier recommendations of the Standing Conference was directed towards chief constables, urging the establishment of crime prevention panels to harness and promote local efforts in the

---

[6] *Crime Reduction*, Annex B, p. iv.

prevention of crime. Since 1968, over 200 such panels have been set up by police forces in England and Wales. Today the Standing Committee provides a national forum for industrialists, representatives from trade and commerce, voluntary groups, and local and central government to review progress and consider discussion papers which are commissioned and published on various aspects of prevention. In 1986 there were five reports: on the prevention of violence associated with licensed premises, on commercial robbery, on burglary, on car crime, and on shop crime. The reports are practical in nature and aim to make specific preventive suggestions which are capable of being acted upon.

The way in which crime prevention panels work in practice varies from one force to another, but typically they comprise a cross-section of local community representatives, such as industrialists, businessmen, teachers, clergymen, representatives of voluntary organizations, and the press. At first the chairman was usually a senior police officer, but there is now a move towards civilian chairmen. Panel members are recruited by personal approaches by the police or on the recommendation of other panel members. The local police force usually provides a small budget, which some panels have been able to supplement with donations from local commercial interests. Sponsorship is increasingly being used as a means of raising funds and widening support for crime prevention. Junior panels and panels in the work-place are among the more recent developments. Activities extend to arranging and manning exhibitions, publishing and distributing leaflets, liaising with the schools, setting up communication networks between retailers alerting them to the presence of shoplifters in the locality, and fitting door chains and viewers at the homes of elderly people free of charge or at a reduced cost. Additional protection of this sort is much appreciated and is sometimes facilitated by donations from commercial interests. The emphasis is on how particular problems can be countered, whenever possible drawing on data made available by the police or victims' support schemes.

Outside the Home Office, other central government depart-ments have shown a willingness to take crime prevention into account when pursuing their own primary objectives. A good example is the crime prevention initiative set up within the

Department of Employment's Community Programme. This is a publicly funded temporary employment scheme, administered by the Manpower Services Commission, offering work of practical benefit to people who have been unemployed for some time. A variety of organizations, including local authorities, voluntary bodies, and private businesses, run schemes under the ambit of the Community Programme. The projects must relate to work that would not otherwise be done and in which any private gain is secondary or incidental to the public benefit. Within the wider goals of the Community Programme, the crime prevention initiative, launched in conjunction with the Home Office in 1985, encourages agencies and sponsors to consider the possibilities of crime prevention. In its first year, projects were found to provide work for over 5,000 unemployed adults, representing funding of the order of some £22 million in a full year.[7] The work was aimed at improving the protection of people or property, or building up social and community activities. The participation and support of local people and organizations was treated as an essential ingredient.[8] Nearly 200 projects covered the provision of twenty-four-hour porters in vulnerable tower blocks, teams fitting locks at the homes of the elderly or disabled, property-marking, help for women victims of domestic violence, and enhancing community youth and leisure activities.

The value of such a scheme is twofold: it supplies a service which helps to reduce crime by limiting the opportunities for offending, while simultaneously occupying in a constructive way the time of some of those who might otherwise be tempted to make use of such opportunities as exist. Young people especially have shown they can be part of the solution as well as part of the problem. Another example that has attracted notice is a large-scale programme initiated and managed by the police in Staffordshire. This enterprising scheme, known as the SPACE programme (Staffordshire Police Activity and Community Enterprise), provides a range of supervised activities for as many

[7] *Criminal Justice—A Working Paper,* p. 9.

[8] One example, quoted by the Home Secretary in addressing the Standing Conference on Crime Prevention on 18 Nov. 1986, was an association between the National Westminster Bank and Age Concern. NatWest provided the necessary hardware to allow a lock-fitting scheme run by Age Concern under the Community Programme in Avon to expand its work fitting locks to the homes of the elderly.

as 23,000 young people between the ages of ten and sixteen during the school summer holiday period. In its present form the scheme dates from July 1983, although the origins go back to 1979. By 1982 it was being operated in all six police divisions. The programme is repeated annually and has earned good support in the area, from local authorities, businesses, and, perhaps most encouraging of all, from over 1,000 members of the public who volunteer their services. The Chief Constable reports that the programme has made a substantial contribution to the reduction of juvenile crime in Staffordshire.

III

It is now time to explore more closely what is meant by the hard-worked expression 'the community' in the context of crime prevention, dissecting the interplay of people and dwellings in the physical environment. Housing is the focus, because it is housing that constitutes the elemental framework of residential communities. While it is true that a considerable volume of crime takes place outside the home, particularly offences of violence against the person, it is domestic security that looms largest in the context of crime prevention. The home is the one place above all else which those living there instinctively want to keep private and safe. One of the inherent problems is that people cannot be protected as easily as their property. The importance of shutting and locking doors and windows at night and when going out is a simple message to communicate and one that is readily understood. But householders cannot be expected to remain at home all day behind locked and bolted doors. Leaving the practicalities of everyday living on one side, the concept of a fortress society is not an appealing one on social grounds. So physical security measures—good locks and chains on stronger doors and frames, window locks, alarms, and other security devices—need to be supplemented by observation and watchfulness. When the householder is at home, he can keep a weather eye not just on his own property, but on that of his neighbour too. Policemen patrolling their beat on foot are more observant than those in cars, and resident caretakers, once a dying breed on housing estates, are returning again as the realization sinks in that the human eye and brain are still the most sensitive and reliable of all warning devices.

In public sector housing, now a declining proportion, although at the end of 1985 still representing 27.3% of the national housing stock,[9] there are special problems. Although local authority housing standards overall are no lower and sometimes higher than those found in other forms of tenure, certain estates were so inappropriately designed, particularly those built in the 1960s and early 1970s, that they embody what the Inquiry into British Housing, set up by the National Federation of Housing Associations and chaired by the Duke of Edinburgh, called 'deficiencies on a grand scale'.[10] The report referred to thousands of council properties which not only were inappropriately designed and badly built, but had proved impossible to manage or maintain properly. Among the more common factors making life a misery for tenants were condensation and inadequate heating systems, bad insulation or noise from adjoining dwellings, structural deficiencies in the building, and design and layout faults contributing to insecurity, vandalism, and fear of crime. The sheer scale of the largest estates—well over 1,000 dwellings on a single site—and their location did nothing to inspire a sense of community in the disparate and not always willing residents allocated to them. Poor communal facilities, such as public transport, shops, churches, and meeting places, and dissatisfaction over maintenance and repairs, added to the disenchantment of the inhabitants.

Nor were the inhabitants themselves inherently stable groups capable of coalescing naturally into cohesive communities. On the contrary: the decline in the private rented sector resulted in a large majority of those who could not afford owner-occupation turning towards local authority housing. Until the mid-1970s the pressure was such that virtually no house or flat was impossible to let, whatever its condition. There were long waiting lists and little choice for prospective tenants, including many families

---

[9] 27.3% represents the proportion of dwellings rented from local authorities or new town corporations over Great Britain as a whole. The comparable figure for England in December 1985 was 25.0%. Department of the Environment, *Housing and Construction Statistics 1975–1985,* HMSO, London, 1986, Table 9.3, p. 98.

[10] *Inquiry into British Housing: Report,* National Federation of Housing Associations, London, 1985, p. 4.

with social problems. Then, however, a new and unexpected phenomenon emerged. This was graphically described by the Director of the Housing Services Advisory Unit at the Department of the Environment in addressing a seminar on crime prevention in 1982:

Certain estates, certain areas, became difficult to let. By this I mean that some estates became so unpopular that applicants for housing would accept accommodation on them only as a last resort. Consequently, only applicants in desperate need, usually those with a host of social problems—the homeless, single-parent families, the unemployed—would accept offers of homes on such estates. The unpopularity of these estates was due mainly to unsuitable location, poor design, a lack of proper maintenance, and crime.

And so, sadly, a spiral of decline set in. Vandalism emerged as a major problem in certain areas and it was found that crime levels were often higher within these problem estates. From about 1972 onwards, more and more councils found they had problems with difficult-to-let estates, and it was obvious that something had to be done to halt the decline ... there are now [in 1982] some 280,000 dwellings classified as difficult-to-let in the United Kingdom: 60% of these were built since 1965, but the total also includes some pre-war housing. One half of this problem housing is situated in metropolitan areas and two-thirds are maisonette-type dwellings.[11]

The last words are important. By no means all of the problems were encompassed in the popular stereotype of ill-constructed high-rise blocks of flats in inner-city areas, some only recently completed, being vandalized by unruly gangs of youths, with the quality of life at a low ebb. These factors were certainly present and contributed to crime and delinquency, but they did not provide a complete explanation. The reality was that, not only was a variety of buildings and surroundings deteriorating rapidly, but many of the people who lived in such housing were demoralized and embittered, antagonistic towards authority, resentful of what they saw as the remote high-handedness of local housing managements, hostile and suspicious of the police, and divided among themselves. Racial tensions added dangerously combustible material in certain areas. As the 'good' tenants demanded to move away from run-down estates, either

---

[11] *Crime Prevention: A Co-ordinated Approach,* Proceedings of a Seminar on Crime Prevention, Police Staff College, Bramshill House, 26–29 September 1982, HMSO, London, 1983, p. 7.

they were replaced by still more problem families, or increasingly the houses or flats were left empty, with windows boarded up and squatters taking over, and with a consequent reduction in the rent and rates which the local authority could collect. In 1981 there were some 24,100 local authority dwellings in England which had been empty for more than a year, adding to homelessness and resulting in substantial rent and rate losses.[12] In addition, there was the cost of securing the dwellings and making good the damage done before trying to re-let them.

Opportunities for vandalism and more serious offending abounded in such conditions. All of the features identified in Chapter 6 as the springs of delinquency were present. The above-average proportion of one-parent families and absent fathers diminished the potential for parental guidance and example. In areas of multiple deprivation schools were unlikely to attract the better teachers, while the malign influence of the peer group was everywhere pervasive. Although it is hard to draw the line between petty offending and conduct regarded as being a nuisance, or uncivilized behaviour between neighbours, it was evident that in some of the difficult-to-let estates reported levels of crime were well above average. In this respect they did not always surpass the record of older estates with settled criminal sub-cultures, but that was little consolation to residents who had only recently arrived and who often wished to move out again as soon as possible.[13] Resistance to external intervention, whether by police, probation, or social work agencies, was underlined by the fact that so much offending or anti-social behaviour (e.g. between neighbours) was internal, being committed within the confines of the estate. Thefts and burglaries were made easier by knowledge of the strong and the weak, the careful and the careless, as well as more immediately by who was out or away from home. Even in the case of physical assaults, there could be an awareness of who the assailant was,

[12] Department of the Environment, *Report on Local Authority Empty Dwellings*, HMSO, London, 1984, Appendix 3. The total of empty dwellings for all local authority housing in England was 97,400 in 1981, rising to 107,100 in 1983.

[13] See A. E. Bottoms and P. Wiles, 'Housing Policy and Crime Reduction Strategies', paper for a Home Office Workshop on Communities and Crime Reduction, Cambridge, July 1986. It is expected that an edited version of the papers will be published by HMSO in 1987.

where he lived, and the risks of victimization if the crime was
reported.

IV

The natural history of the problem estates invited study, and the
results of study compelled attention from more than one quarter.
Both central government and local authorities were appalled by
the prospect of a massive public sector investment deteriorating
at a rate that might make it necessary to pull down relatively
recent housing before the loans raised for its construction had
been repaid.[14] Some housing, regarded as uninhabitable, was in
fact demolished, while other estates were sold to private builders
or developers for refurbishing. Meanwhile, in 1979 the
Department of the Environment combined with local housing
authorities to launch the Priority Estates Project, with a view to
seeing what could be done by improving the management of
difficult-to-let estates, decentralizing responsibility for repairs
and maintenance, improving lighting and security, and making a
determined effort to build up closer links with the residents. The
original aims were described by the Department's consultant as
being:

to establish a local management office, to carry out meticulously the
landlord's responsibility for rent, repairs, letting property and
maintaining the environment of the estate; and to give tenants a chance
to exercise maximum control over their homes and neighbourhood.[15]

Five years later, the same author was able to add:

It seemed clear from the outset that only by harnessing the energy and
goodwill of residents and workers in a local organization could the
problem of the run-down, difficult-to-manage estates be reversed. Many
councils have done just that.[16]

It is impossible to be precise in judging the effects of improved
housing management on criminality and offending. Although the

---

[14] *Crime Prevention: A Co-ordinated Approach*, p. 8.
[15] Department of the Environment, *Local Housing Management: a Priority
Estates Project Survey*, HMSO, London, 1984, p. 1. The consultant was Anne
Power.
[16] Ibid.

Priority Estates Project was not aimed directly at crime reduction, the Department of the Environment had become aware of the significance of vandalism and the fear of crime on the problem estates. The projects attracted wide interest, from criminologists as well as those with an interest in housing policies. It also complemented some existing initiatives taken by the National Association for the Care and Resettlement of Offenders (NACRO) with local authority support. NACRO's approach was more specifically concerned with crime, seeking to confront vandalism and minor crime by involving the residents on the most difficult estates in planning improvements in such a way that they would feel inclined to maintain and protect them.[17] Starting at an estate at Widnes, Cheshire, in 1976, NACRO broadened its range with the establishment of a Crime Prevention Unit in 1979 and a Safe Neighbourhoods Unit in 1980. By 1986 there were sixty NACRO Crime Prevention Units working in forty local authority areas, and Lord Donaldson of Kingsbridge, the NACRO president, was able to report some encouraging results in a House of Lords debate:

The Widnes unit, when it was first set up in 1976, revealed vandalism to one in every three houses. By the 1979 survey, this was reduced to one in every eight. In 1976, 70 per cent of the residents reported a burglary against 8 per cent in 1979. Police and survey workers both reported good progress in reducing crime. Later monitoring has confirmed these results. To take another example, this time in the London area, in the nine months to June 1985, compared with a similar period to June 1984, on a Lewisham estate burglaries had dropped by 54 per cent, car crime by 60 per cent and street crime by 48 per cent. These are just two instances; but I am talking about something which is not difficult to do, not expensive to do, and is working.[18]

NACRO received further support from public funds in December 1986, when the Government agreed to finance a NACRO initiative known as the Inner-cities Crime Prevention Development Unit as part of its own policies towards tackling

[17] Professor Paul Rock's paper, entitled 'Research on Crime Reduction Initiatives on Problem Estates', prepared for the Home Office Workshop on Communities and Crime Reduction, describes the NACRO and Priority Estates Project initiatives in some detail. Rock concludes that 'Their work is of major interest because it promises to diminish rates of crime and make unpleasant conditions more tolerable.'

[18] *Parl. Debates,* HL, 471 (5th ser.), cols. 1070–1, 26 Feb. 1986.

the problems of employment and disadvantage faced by inner-city residents and alienated young people. Five full-time staff from the NACRO unit were to work alongside public officials in eight selected areas to devise and implement programmes designed to reduce crime and the fear of crime. At the same time, a consortium formed by the Apex Trust, the National Youth Bureau, and the Intermediate Treatment Fund also received financial support.[19]

In the private housing sector, the social disabilities do not apply with the same intensity as on the problem council estates, but the morals—a closer attention to housing management, the importance of landlord–tenant relationships, and the effective delivery of services—are equally relevant in housing rented from private owners. Irrespective of ownership or tenure, as neighbourhoods go into decline those who remain become less willing and less able to control anti-social activities, breeding a fatalistic attitude towards criminal behaviour. Such tolerance does not seem to lead to any greater security. Indeed, the data from the British Crime Survey, which is shown in Table 18, indicates that, with the exception of the best-quality houses, flats, offices, and shops in the city centres (classified as 'high-status non-family areas'), the poorer the neighbourhood, the greater the risk of being victimized by crime.[20] The geographic concentration is strengthened by the fact that the high-status, non-family areas are often adjacent to, or even contain, pockets of social disadvantage.

In terms of risk of victimization, owner-occupiers are sometimes, but not always, more favourably placed than local authority tenants. Some research has been done on the ecology of tenure which suggests that tenants in the poorest council estates face a risk of burglary some five times greater than the national average. That this is more to do with specific environments than with the forms of tenure is shown by the finding that just over half of all council tenants in England and Wales face a risk of burglary which is equal to or less than the national

---

[19] *Parl. Debates*, HC, 107 (6th ser.), col. 542, 17 Dec. 1986.
[20] Table 18 was prepared by T. Hope and M. Hough of the Home Office Research and Planning Unit for the Workshop on Communities and Crime Reduction. Crown copyright; reproduced by permission.

TABLE 18. *Crime rates, by risk areas, 1984*

	Burglary incl. attempts (% households)	Robbery & theft from person (% persons)	Autocrime outside the home (% owners)
Low-risk areas			
Agricultural areas	1	1.3	5
Older housing of intermediate status	2	0.8	10
Better-off retirement areas	3	1.1	6
Affluent suburban housing	3	1.1	7
Modern family housing, high incomes	3	0.9	8
Medium-risk areas			
Better-off council estates	4	1.4	14
Poor-quality older terraced housing	4	1.4	18
Less well-off council estates	4	1.4	15
High-risk areas			
High-status non-family areas	10	3.9	15
Multi-racial areas	10	4.3	26
Poorest council estates	12	3.3	21
National average	4	1.4	11

average of about 4 per cent.[21] Nevertheless, owner-occupation offers some advantages when it comes to policies for crime reduction. Typically, the householder will have chosen where he wants to live, in a house and area he can afford, rather than having been allocated to it by a bureaucratic procedure.

[21] T. Hope, 'Council Tenants and Crime', Home Office, *Research Bulletin* 21, 1986.

Professor Anthony Bottoms has argued that the way in which housing is allocated—i.e. by a market price mechanism or by bureaucratic allocation—will have an effect on crime rates within the residential community.[22] The system of housing allocation will determine the mix of residents in an area and the social life they create. It is not only age, occupation, family size, income, and ethnic origin that will differ, but also the relative crimino-genic potential of residents. This has been shown quite clearly when complete and thriving criminal sub-cultures have been transplanted into new housing estates as a consequence of slum clearance programmes.[23] Less obviously, if vandalism is committed primarily by children and younger teenagers (as empirical research suggests), then an allocation process which concentrates in certain areas families with a larger number of younger children, and looser social controls on their behaviour, will result in a higher rate of vandalism in those areas.

V

This analysis suggests that social and environmental factors interact in forming the community and determining the levels of offending within it. It follows that a strong and self-reliant community, with pride of place and ways of controlling deviant behaviour, will be more capable of preventing crime than a divided and demoralized one. This is simultaneously the justification for the emphasis on community development and the explanation of why neighbourhood watch and other self-help schemes have initially taken root in the lower-risk areas with a greater degree of owner-occupation. Unlike victims' support, where volunteers working in deprived and problem areas may not themselves live in those areas, neighbourhood watch is intrinsically bound up with those who do live or work in the neighbourhood.

The rapid growth of neighbourhood watch to a record total of 17,500 separate schemes in England and Wales by December

[22] Bottoms and Wiles, 'Housing Policy and Crime Reduction Strategies', pp. 2–5.
[23] Ibid.

1986, a number that had more than trebled in just over the previous twelve months, owes little to policy initiatives by central or local government. It has been a manifestation from the grass roots, originally seeded by police in Cheshire and Avon drawing on some American experience,[24] but then taking off spontaneously. The original idea was a simple one, and it struck a responsive chord in many communities throughout the country. Although there were local variations, the basic principle was that residents living in a specified area should organize themselves in such as way as to forge effective links with the police for the better protection of their property. A concentration on improved standards of physical security (locks, bolts, and alarms) was backed up by a system of informal surveillance, reporting anything untoward or suspicious to the police. The geographical expansion of neighbourhood watch has been accompanied by an enlargement of the concept to cover the strengthening of communal bonds, a closer and more mutually appreciative relationship with the police, and the creation of habits and attitudes hostile to crime.

With these added values, the next step must be for neighbourhood watch to branch out from the fertile soil of the more stable and prosperous areas where it has flourished so spectacularly and to succeed in penetrating the crime-prone housing estates where, as we have seen, the need and the potential for crime reduction is even greater. Whether neighbourhood watch can survive in less homogeneous communities, reluctant to co-operate with the police, or will be rejected as an alien implant is a question that goes to the heart of prevention policies. Local authorities have an especial responsibility towards those of their tenants who live on high-risk estates with little opportunity to move away. It is not just the housing aspects that need to be remedied, or the management neglect and design faults: in some cases it is political attitudes that are impeding progress. Certain councils, for example, have taken an obstructive line with

---

[24] Surveys in the United States indicate that 5% of the American public had participated in neighbourhood watch in 1981, rising to 7% in 1984. Almost 38% of householders participated in areas where programmes existed. See D. P. Rosenbaum, 'A Critical Eye on Neighbourhood Watch: Does It Reduce Crime and Fear?', paper for the Home Office Workshop on Communities and Crime Reduction, 1986, p. 13.

residents wishing to set up neighbourhood watch schemes, even to the point of refusing planning consent for the street signs.[25]

The high visibility of neighbourhood watch, with street signs and window stickers warning would-be criminals to steer clear of localities where the risks of detection and arrest are greater than elsewhere, raises the long shadow of displacement. Put briefly, there will not be a net gain in the overall protection of society if what happens is that potential burglars or other offenders are simply moved on from a harder target to a softer one. This possibility clearly exists, and the issue is an important one to try and resolve; for if it can be established that substantial displacement does occur, then those who have invested so heavily in resources and effort in situational crime prevention will have to look elsewhere in their attempts to achieve a reduction in the volume of criminal offending.[26] Alternatively, and this seems the more logical option, the existence of displacement might be taken as an indication that crime prevention measures can have an effect on crime, and that it is wrong to assume that crime, like the weather, is beyond our control. In this context, the existence of displacement suggests the need for more preventive measures and not fewer. That said, displacement is an elusive concept, which is not surprising since it is the motivation and degree of determination of the potential offender that will govern whether or not he tries again. Some, with stronger intentions and perhaps greater skills acquired from previous experience, may adhere to their intention to burgle a dwelling, switching to another target or date if they are frustrated initially. But many other crimes are committed more casually, under conditions of stress or after heavy drinking, or with encouragement by a peer group. Impulsive crimes, carried out on the spur of the moment, are facilitated by easy opportunities—unlocked cars and unprotected

---

[25] Five London boroughs (Greenwich, Newham, Haringey, Hackney, and Lambeth) and Manchester were named in a Parliamentary Answer which disclosed that the Department of the Environment was considering whether deemed consent could be given to neighbourhood watch street signs, thus removing the need for planning permission and planning fees. *Parl. Debates,* (HL), 484 (5th ser.), col. 4, 2 Feb. 1987.

[26] K. Heal and G. Laycock, 'Principles, Issues and Further Action', *Situational Crime Prevention: From Theory into Practice,* pp. 123–4. See also P. Mayhew, R. V. G. Clarke, A. Sturman, and J. M. Hough, *Crime as Opportunity* (Home Office Research Study no. 34), HMSO, London, 1976, pp. 5–6.

homes. To the extent that these opportunities can be reduced, then it should follow that property offences which are essentially opportunistic will decline. Only the most tentative conclusions can be put forward in the present state of knowledge. The editors of a recent Home Office research publication tracing the evolution from theory to practice of situational crime prevention put it like this:

displacement is more likely to take place where the individual's motivation is sufficiently high to drive him on even when his initial target of criminal activity is well defended. Where his motivation is lower the protection of the target may well be sufficient to deflect the potential offender from crime altogether ... On the basis of these arguments it would seem reasonable to suggest that situational prevention can reduce crime by influencing the final decision of some potential offenders, and that even where displacement takes place, only a proportion of the initial potential offenders will pursue their intent to commit crime.[27]

VI

Crime prevention now faces some difficult questions of funding. Both the Association of Municipal Authorities and a report by the working group on residential burglary at the second of the Downing Street seminars estimated the cost of a selective grant scheme for elderly and especially disadvantaged inner-city residents at between £60 and £65 million. With large numbers of properties still standing empty on unpopular or run-down council estates,[28] the Department of the Environment continued with its Priority Estates Project aimed at strengthening and localizing housing management in twenty-five of the most difficult-to-let estates. In November 1986 the Minister for Housing, Urban Affairs, and Construction, John Patten, announced further expenditure of £1.78 million to improve security in 2,835 homes on London council estates with high rates of crime. The specially constructed improvement packages included entry-phones, new doors and locks, improved lighting,

---

[27] Heal and Laycock, *Situational Crime Prevention*, pp. 124–5.
[28] Local authority dwellings may be empty for a number of reasons. Some are vacant and available for let, while others are undergoing repairs or improvement, or awaiting repairs or improvement, or awaiting sale or demolition.

and the blocking off of dangerous communal areas. This addition was part of a programme launched in June 1985 under which the Department of the Environment's Urban Housing Renewal Unit visited 120 local authorities in England and agreed on proposals with them for a range of measures, including joint ventures with private developers and local estate-based management on the lines pioneered by the Priority Estates Project. Support under this programme amounted to £41 million in the first half of the financial year 1986/7. Nearly 45,000 homes have benefited.[29] Despite these spending programmes and improvement schemes, resulting in part from the heightened publicity generated by crime prevention, much scope remains. Some residents in low-cost private rented accommodation in particular are at greater risk than necessary, since they may not have the means to pay for improved security, whereas their landlords may be unwilling to do so.

For the growing number of owner-occupiers,[30] there are financial incentives in tax relief on loans made for home security purposes,[31] and the prospect of premium discounts of between 5 and 15% to holders of household contents insurance policies who improve the security of their homes. Building societies too are showing an interest in crime prevention, and in 1986, at the instigation of the Building Societies Association, representatives of the ten largest societies met a Home Office Minister and officials to discuss positive steps. At this meeting it was agreed that, because of their widespread involvement in the domestic housing market, societies' efforts could best be directed towards increasing public awareness of the need for and benefits of adequate home security as the best means of burglary

[29] Department of the Environment, *News Release,* 24 Nov. 1986.

[30] Owner-occupied dwellings, expressed as a percentage of the total housing stock in Great Britain, rose from 55.3% in December 1979 to 61.9% in December 1985. Department of the Environment, *Housing and Construction Statistics 1975–1985,* p. 98.

[31] In October 1986 the Building Societies Association advised its members that, although the Inland Revenue had always recognized that a loan for the installation of a burglar alarm qualified for tax relief on the interest payments, there had been some doubt as to whether the fitting or replacement of security locks qualified as a home improvement for the purposes of obtaining tax relief. In August 1986 the Inland Revenue confirmed that 'the fitting or installation of anti-theft devices such as burglar alarms, security locks for windows and doors and warning light systems is considered to be a qualifying improvement for the purposes of Part I of Schedule 9 to the Finance Act 1972'.

prevention. Building societies have a close relationship with large numbers of prospective home-owners as well as with those who already hold a mortgage. They are ideally placed to direct attention to the benefits of home security at a time when borrowers are most receptive to the idea of protecting their homes against the possibility of burglary. Societies can provide additional finance should it be needed to enable the borrower to fit locks, alarms, and other security devices, either at the time the first mortgage is made or subsequently in further advances for home improvements. Through their network of High Street branches, building societies also have an enormous potential for the distribution and display of printed publicity material.

At this stage in the progress of crime prevention, it is sensible to avoid making sweeping or exaggerated claims as to effectiveness, particularly since national crime figures are not directly related to local crime prevention activities. Nevertheless, there are some encouraging straws in the wind, locally as well as nationally. In its divisional annual report placed in libraries and other public buildings, the Metropolitan Police in the southern half of the London Borough of Islington recorded a reduction in all reported crime of 16% in 1985 compared with the previous year. The most prevalent crimes of burglary and autocrime (theft of or from a motor vehicle) were reduced by 22% and 24%, respectively.[32] The police attributed the reduction to more uniformed officers patrolling the area and a significant increase in community initiatives, especially neighbourhood watch. By 1 December 1986 the number of watch schemes had grown to twenty-nine, covering some 24,000 people out of a total resident population in the police division of about 90,000. Nationally, too, one of the few downward trends reported in the 1985 _Criminal Statistics_ was a decrease of 4% in the number of burglaries from dwellings in 1985. This compared with a rise of 10% in the previous year, and an average annual rate of increase of 5% in the ten years since 1975.[33] Part of the recorded growth in domestic burglaries over the previous decade was attributable to an increase in the proportion of offences notified to the police

[32] Metropolitan Police (King's Cross Road Division), _Divisional Annual Report 1986_, p. 3.
[33] _Criminal Statistics England and Wales, 1985_ (Cm. 10), HMSO, London, 1986, p. 22.

and recorded as burglaries. The British Crime Survey also disclosed the substantial number of attempted burglaries which were unsuccessful because the burglar was unable to gain entry, or was disturbed by occupants, neighbours, or passers-by.[34] If present trends continue, there seem to be adequate grounds for believing that the practical expression of the public's desire to protect itself with the weapon of crime prevention is proving effective.

VII

From the prevention or reduction of crimes against the home or possessions to the prevention or reduction of violent crimes against the person is a giant leap. Are the techniques of situational crime prevention which have made property offences harder to accomplish capable of being adapted to safeguard the individual who is at risk of suffering personal injury while away from the home? It is not simply the chances of being mugged or assaulted in the street or other public places that have to be assessed. There are risks for passers-by as well as customers outside rowdy pubs in city centres at closing time. There are risks in late night travel on buses, trains, or undergrounds, and staff are at risk in an unexpectedly wide range of occupations when at their place of work. What, if anything, can crime prevention do to reduce these risks? In considering these questions, it is as well to return to the first principles stated earlier in this chapter. The aim of situational crime prevention is to seek to reduce the potential for crime through the management, design, or manipulation of the immediate environment in which it occurs. Unlike burglary and theft, which are widespread, non-domestic crimes of violence tend to be more specifically linked to particular settings.

A good example can be found in crimes associated with premises licensed for the sale and consumption of alcohol. The licensing regulations constitute the primary controls, principally on the hours during which alcohol can be sold to the public, and

[34] See M. Hough and J. Mo, 'If At First You Don't Succeed: BCS Findings on Attempted Burglary', Home Office, *Research Bulletin,* 21, 1986, pp. 10–13.

the police are the usual enforcement agency. But within this framework, there is scope for preventive action. Incidents of disorder and violence take place in and around licensed premises—hotels, restaurants, clubs, public houses, and off-licences. One by-product of the licensing laws is to concentrate violent incidents into a comparatively short period of time, immediately after the pubs and clubs close their doors. Surveys in the northeast of England indicate that something like a quarter of all city-centre incidents of violence reported to the police between 6.00 pm and 6.00 am the following morning take place within half an hour of the pubs' closing time, when their customers are decanted on to the streets, with a further although smaller peak of incidents at 2.00 am, when the clubs close.[35] Flexible hours of opening would be one response to the problem, but that might have the effect simply of moving on the most unruly and determined drinkers from one pub to another until the last port of call had closed.

Overcrowded and noisy pubs in city centres are difficult to supervise, and they compete with one another to attract a transient and largely youthful trade. Heavy drinking and loutish or aggressive behaviour drive away the more stable regulars, and without their steadying influence a volatile and mobile clientele is left. Most of the larger premises are directly managed by the companies who own them rather than being tenanted. The pubs that are the most difficult to manage typically have the highest turnover of staff and managers, with the paradoxical result that the most inexperienced managers may be placed in the most troublesome pubs.[36] Away from the city centres, the national picture obtained from British Crime Survey data and police forces is of perpetrators of violence being in an age range between eighteen and thirty, mainly male, semi-skilled or unskilled workers, or unemployed. The victims too are largely male, in the same age range, skilled manual or unskilled workers

[35] T. Hope, 'Liquor Licensing and Crime Prevention', Home Office *Research Bulletin,* 20, 1986, pp. 5–12. The Newcastle study was based on telephone calls made to the police from licensed premises and related to incidents of drunkenness, disorder, assault, and criminal damage in public places on Friday and Saturday nights in the city centre of Newcastle upon Tyne. One particular trouble spot, no more than 250 square yards in area, contained twelve public houses.

[36] Ibid., p. 7.

in the habit of going out four or more evenings a week. They were more likely to be lonely men—single, divorced, or separated—and to be either unemployed or in part-time employment only.[37]

The trade interests have a good reason to dislike alcohol abuse and rowdyism: it is their employees who bear the brunt of threats and violence, and their valued customers who are driven away. Thus the brewers, the licensed victuallers, and the licensed house managers joined Home Office officials and representatives of local government, the police, and the probation service in a working group set up by the Standing Conference on Crime Prevention under the chairmanship of a prominent brewer and magistrate.[38] In November 1986 the working group submitted a grim report to the Standing Conference on the nature and extent of violence and disorder associated with licensed premises. It depicted frequent incidents of fights breaking out after quarrels (noting an increased use of weapons), and occasional pitched battles between rival gangs, sometimes but not always related to football. Smaller groups of younger people mixing amphetamines and other drugs with alcohol could 'go berserk', while under-age drinkers were generally thought by the police to be widely involved in violence and disorder both inside and outside premises. Less obvious were incidents of bar staff being threatened for free drinks or forced to give change for higher-denomination notes than those offered in payment for drinks. Those involved in this type of activity were in the thirty-to-fifty age range, hard men, often long-term unemployed and with criminal records.[39] Whatever the cause, violence was prone to spill over into the street, and physical injuries were more likely to result outside the pubs than inside.

The report concluded that a pro-active approach was required towards disorder and violence in licensed premises rather than the traditional re-active approach relying almost exclusively on legislation, arrest, and sentencing for control.[40] Education

[37] Home Office Standing Conference on Crime Prevention, *Report of the Working Group on the Prevention of Violence Associated with Licensed Premises,* Home Office, London, November 1986, p. 3.

[38] Ewart A. Boddington, JP, Chairman of the Brewers' Society, 1984–5.

[39] Home Office Standing Conference on Crime Prevention, p. 8.

[40] Ibid., p. 15.

programmes and responsible parenting, a familiar theme, were seen as the best way of warning young people of the dangers of excessive consumption of alcohol. A public education campaign, on at least the scale of the Road Safety programme, was called for. Breweries were recommended to give more consideration to the establishment of discos, or 'fun pubs' for young people, serving either alcohol-free drinks or drinks with limited alcoholic content. Licensees and bar staff needed improved training on the enforcement of the licensing laws and ways of handling difficult customers. An investigation of best practice in the maintenance of order was recommended, with the results incorporated into training programmes. Reminiscent of crime prevention on the problem housing estates was the recommendation that, 'in those establishments which are notoriously difficult to operate without problems, consideration be given to their establishment as "community pubs", perhaps guided by a local residents' group'.[41] Police were urged to move away from their role as disinterested enforcers of the law to co-operate with the other interested parties—the brewers, the licencees, and the local authorities—in attempting to pre-empt potential offending. The report was well received and widely circulated. It is filled with good sense on a matter of urgent public importance, and it is to be hoped it will concentrate attention on what needs to be done, how it can be done, and by whom.

Local police in some areas have already shown themselves ready to take up the challenge of unruly licensed premises and the violence they generate. The same police division in north London that reported a drop in property offences in 1985 is now giving priority to violent crime.[42] Licensed trade interests, magistrates, and the local council have come together in a joint effort to counter disorders connected with licensed premises which over a recent three-month period had led to one murder, four threats to kill, and twelve grievous bodily harms, as well as robberies, offences of arson, criminal damage, and the possession of dangerous weapons. Drug dealing, under-age drinking, and rowdyism associated with football supporters were additional illegal or anti-social activities not reported to, or

[41] Home Office Standing Conference on Crime Prevention, p. 17.
[42] 'N' Division (King's Cross Road) of the Metropolitan Police, covering the southern half of the London Borough of Islington.

recorded by, the police. All 250 publicans and other licencees in the police division were circulated, warned of enforcement action by plain clothes and uniformed officers, and invited to attend meetings to discuss action along the lines recommended in the Standing Conference's report.

Places of work can be as dangerous as places of recreation. Trade unions and employers, in the public service and in the private sector, have become increasingly concerned in recent years about the risks to staff whose work brings them into direct contact with the public.[43] Risks vary from threats and abuse to physical assaults, sometimes leading to permanently disabling injuries or even, in rare cases, to loss of life. It is doubly tragic when people who are engaged in providing a service to the public, such as bus drivers or staff in social security offices or hospitals, are abused or attacked. Assaults may result from criminally inspired attempts to rob or defraud, but many are caused by belligerence and aggression which get out of control as a consequence of emotional distress, frustration, or sheer despair.[44] Some members of the public will be suffering from mental disorders, while others inevitably will be under the influence of drugs or alcohol, the latter especially presenting endemic problems to passengers as well as staff on late-night public transport.

The situations in which working staffs are at risk call for continuous monitoring to devise and implement preventive strategies for their greater safety. Some will be managerial—for example, reducing waiting times that have such an adverse effect on tolerance, and preventing crowds from building up. Staff redeployment may be successful if difficult social security clients or applicants for local authority housing are steered towards more experienced staff. In some cases face-to-face contact may be avoided altogether. Examples are the adoption of automatic ticketing on transport systems thus eliminating personal contacts with ticket collectors and inspectors, and the payment of rent by

---

[43] In 1986 the Health and Safety Executive published a survey by the Tavistock Institute of Human Relations which had been commissioned by an interdepartmental committee set up to study the problem of violence to staff. The report provided a framework for analysis and outlined the scope for positive action in a number of settings. See B. Poyner and C. Warne, *Violence to Staff: A basis for Assessment and Prevention*, HMSO, London, 1986.

[44] Ibid., p.3.

non-cash transactions.[45] Where cash is held improved security precautions include protective screens for counter staffs in banks, post offices, and building society branches, and also at filling stations, where staff working on their own at night can be targets for robbery. Tense or explosive situations can be prevented from escalating into assaults by the installation of discreet emergency buttons in housing department interview cubicles. Out of an additional £15 million being provided by the Government over three years to combat crime on the London Underground, direct radio links to the police will be installed at forty-two stations with an above-average record of robberies and assaults, and the station improvement programme will be extended to incorporate passenger alarms on platforms and better lighting in tunnels along which pedestrians walk.[46] Such apparently mundane matters as the width of tables in interview cubicles and the height of the bars in public houses can also be relevant in altering the physical relationship between employee and potential assailant to the advantage of the former.

In each situation, clear procedures need to be evolved, with the co-operation of staff representatives, so that all staff know what is expected of them and how to react if faced with the threat of violence. A sense of proportion helps; out of 15,300 recorded crimes on the London Underground in 1985 (of which theft was the most common), about 1,600 involved violence. This total relates to 725 million passenger journeys made in the year.[47] Hospitals, remarkably, request more police aid than any other single institution, even the pubs and clubs.[48] While incidents include attacks and sexual assaults on staff by relatives and patients, the majority of offences are non-violent. Nevertheless, the fact that one in thirty hospital workers questioned in a survey carried out in 1983/4 admitted to having been a victim of

[45] Poyner and Warne, op. cit., p. 12.

[46] *The Times*, 29 Nov. 1986, reported comments by the Chairman and Chief Executive of the London Underground following the publication of a study by the Department of Transport, in conjunction with the London Underground, the Home Office, the Metropolitan Police, and the British Transport Police, on *Crime on the London Underground*, HMSO, London, 1986. Details were also given in a reply to a Parliamentary Question in the House of Lords: *Parl. Debates*, HL, 483 (5th ser.), cols. 275–6, 18 Dec. 1986.

[47] *Crime on the London Underground*, p. 59.

[48] L. Smith, 'Crime in Hospitals: Some Preventive Possibilities', Home Office, *Research Bulletin*, 22, 1986, p. 24.

personal attack or threat of such an attack, and one in seventy-four of being the victim of a sexual assault or harassment, shows that the risks to hospital staff cannot be disregarded.[49]

From this brief summary of violent crime and the settings in which it commonly occurs outside the home, it is evident that there is scope for preventive measures and that action is already being taken in housing, the licensed trade, public transport, the social services, banks, post offices, and retail outlets. The pressure must be maintained. Some preventive measures are costly, while others depend more on good practice and applied common sense. On the whole, the picture that is emerging is one of positive response, a hardening determination to protect personal safety as well as property by all the means available. Heartening as it is, what we have seen so far is only a beginning, although it is enough to show what can be accomplished. The priority now is to ensure that the requirements of crime prevention are taken into account in the planning and design of buildings and services *ab initio*. People need to be more conscious of crime, and to recognize that with forethought it can be effectively pre-empted by members of the general public, acting singly or in groups. Crime is not inevitable, nor is it simply something to be left to the police and the courts to deal with. The risks of detection and punishment may still be deterrents to some offenders, but prevention is a protection against them all.

[49] Ibid., p. 25.

10

Responses to Crime

I

The framer of penal policy today needs to steer a course between the twin perils of nihilism (nothing that has been tried works) and unwarranted optimism (halve the prison capacity and hope for the best). Neither is a sound guide, although both have articulate adherents. The policy implications of the recent developments that have been described in the preceding chapters amount to a quiet revolution in official and academic assumptions about crime and how it can be contained. Thinking harder about crime—its baffling nature, the situations in which it occurs, and the contrasting types of offender—has led to a broadening of policy responses. Some of what has been happening over the last decade has been obscured by the battle language of the war against crime which has dominated the public debate. In so far as warlike phrases imply resolution and a determination to get to grips with one of the most pressing social evils of the day, all well and good. But too often the rhetoric and slogans are an indulgence, the comfort of opinion without the discomfort of thought, as President Kennedy once reminded a Yale commencement ceremony.[1] The reality is that most crime is committed within communities and among people who have somehow to go on living together, not against communities by enemies who can be defeated or expelled. As the punitive cannons have been thundering away, behind the shot and shell the dispositions towards dealing with crime have been regrouped.

Victims now have achieved a recognized place in the criminal justice system, no longer excluded by the idea that the wrong is against the state rather than the individual who has suffered

---

[1] 11 June 1962. In this speech President Kennedy analysed some distinctions between myth and reality in politics.

harm or loss. Compensation, mediation, and reparation are all relative innovations which allow ample room for enlargement in the future. Professional criminals, the only deviant group of whom it is appropriate to speak in terms of military campaigns, are being brought to justice by information provided by their erstwhile associates as well as by the efforts of the police, who may be unable otherwise to obtain enough evidence to justify a prosecution likely to result in a conviction. Forfeiture and confiscation of the proceeds of drug trafficking and other large-scale organized crime make the potential rewards less attractive, while the greater priority being given to the investigation and prosecution of financial fraud is a change that is closely in tune with public opinion. Non-custodial penalties for the less serious crimes ought not to be seen as soft options. Community service orders are being used on an extensive scale, particularly for younger offenders, diverting from prison considerable numbers of those who are most likely to be contaminated by it; while police cautions for minor offences can often keep first-time offenders out of the embrace of the penal system altogether. This is especially important for juveniles, the youngest group of all, for whom some form of offending is a common (and usually brief) episode in adolescent life. Finally, each year several thousand convicted prisoners are released under supervision as part of a deliberate policy aimed at reducing the incidence, frequency, and gravity of further offending. Some fail to make use of the opportunity offered to them, but the proportion whose licences are revoked because of failure to observe the conditions, or because of further offences committed while on parole, remains at a low level.[2]

Crime prevention holds great promise. It can reduce the opportunities and temptations to offend, thus protecting property and the person. But preventive strategies are also directed towards the motivation of those who might offend and the influences to which they respond. Here the ground is still

[2] Out of a total of 14,406 prisoners released on licence during 1985, 468 licences were revoked for breach of conditions, with an additional 335 being revoked as a result of further offences committed while on parole. All were recalled to prison unless already remanded in custody. As the *Report of the Parole Board 1985* pointed out, some others will have been charged with offences while on parole, but brought to trial only after the licence period has expired. (HC, 428), HMSO, London, p. 3.

largely unchartered and the conclusions speculative—not unexpectedly so, since the terrain to be covered includes the physical and social consequence of alcohol and drug abuse, the tensions of deprived areas and multi-social communities, and the conventions and moral values of a restless nation. While the capability of society to reverse the upward trend of criminal offending with the instruments at its disposal is constrained both by the need to preserve the rights of suspected but unconvicted wrongdoers and by engrained social and moral attitudes, nevertheless, crime can be controlled to some extent by the intelligent application of what has been learned by the study of those who are at risk of offending as well as those who have perpetrated crime, and of the situations in which it occurs.

To some readers the references to 'responses' that are sprinkled throughout the book, if they have caught the eye at all, may have the ring of a modish catch-phrase, fair enough as a title perhaps, but devoid of any greater significance. In this final chapter I would like to argue, on the contrary, that the various forms of response constitute a useful and necessary part of democratic government, acting as a connecting link between the operation of criminal justice and the forces of opinion. Although the word is in common use, response is not easy to define. It is not quite the same as a reply, which implies more of a direct statement or action directed towards another person in answer to a question or request. Responses are provoked by stimuli, events and reports of events, as well as by statements or questions. Public responses may be manifest even when they have not been sought by anyone else, but come to notice because of an instinctive reaction shared simultaneously by large numbers of people. We are not concerned here with the reactions of individuals felt or expressed in private, only with the collective expression and observation of reaction. In the sense of observable public reactions to events or situations, responses are the original sources from which public opinion springs. They are given coherence and direction by organization—by the politicians, propagandists, and lobbyists who keep a watchful eye open to detect the first signs. Once harnessed, the preliminary inchoate responses take on the properties of organized opinion. As such they are acknowledged, however grudgingly, as forming one of the most potent of all political forces.

II

Crime provokes strong responses for a number of reasons. It is immediate and intensely human. At first sight, the majority of criminal offences appear to call for little explanation or exposition: the events seeming to speak for themselves. Crimes can be sensational, shocking, and salacious, or merely pathetic. But they compel attention, more often evoking positive reaction than indifference. The way crimes are reported is all-important, since public rejection of certain forms of deviant behaviour is inherent in the notion of criminality. Every newspaper editor has a sixth sense of what he thinks his readers want to read about, and crime is very near to the top of his list. And not only in the popular press: local newspapers, radio, television, and the quality nationals all carry news of the commission of criminal offences, their detection, and prosecution, virtually on a daily basis. Thus a constant flow of stimuli is maintained, with an awareness of the human interest and a high degree of readership response feeding back into the editorial process, so preserving the supply of stories about crime. In the popular press, or tabloids, there is a resistance to abstract issues and a strong bias towards human interest stories. Crime is straightforward to report and tempting to portray in lurid colours. Agency reports and local stringers can provide news on routine cases, with specialist reporters following up the more important or sensational incidents. The information is generally easy enough to obtain, and provided payments are not made to those who are involved in the crime, relatively inexpensive.

The principal objection is not the fact or even the volume of crime reporting, but the manner. The whole diversity of human character and situation is forced into the mould of an easily recognizable stereotype, the offence and the offender being described in a simplified and repetitive vocabulary. A regular cast of inhuman beasts, monsters, sex fiends, and maniacs is to be found on parade most mornings in the pages of the tabloids, with scant reference to the context in which the crimes took place. Balanced accounts may not be regarded as having a suitably declamatory tone, but the standardized formula employed by the popular press, usually unaccompanied by any explanation or interpretation, demeans the reporting of crime. It is true of course, that crimes, particularly brutal ones, often arouse feelings of revulsion, and that there is a legitimate demand for information by readers. Falling back on this reasoning, editors can claim to be doing no more than reflecting the degree of public interest and satisfying natural curiosity. But the classic justifications for the press—vigilance, enquiry, and disclosure—do not answer convincingly the charges of sensationalism, inaccuracy, and invasions of privacy that bedevil crime reporting. Even though it is sometimes argued that such excesses are the price of a free press, however distressing they may be to the high-minded, this reply cannot excuse the perennial newspaper practice of offering money and actually making payments, sometimes substantial ones, to persons engaged in crime or their relatives or associates.

Competitive pressures of circulation and the satisfying of what are seen as readers' expectations can distort the processes of criminal justice. Witnesses may be suborned, and victims embarrassed and humiliated by thoughtless or unscrupulous reporting. When giving details of a crime to the police, victims do not always realize that their name and address may be given in open court and reported in the press. The subsequent publicity, local or national, can be unexpected and unwelcome, leaving victims with the feeling that they—the ones who have suffered—are being exploited for purposes that can do nothing to help them, so deepening the sense of loss or grievance. Submissions in mitigation made by defence counsel in the course of a trial may have a damaging effect on the reputation of a victim who has no voice in the court.[3] Unchallenged statements referring to a history of mental instability or sexual promiscuity, for example, can get into circulation, causing resentment and distress. Fear of publicity may even deter some victims, especially women victims of sexual assaults, from reporting their ordeal to the police at all. Exaggerated press coverage can also project and reinforce the fear of

[3] In a Statement of Principle on third parties in courts made in 1978, the Press Council said: 'The responsibility for protecting third parties who have no opportunity to reply from unfair allegations and from the publicity which can flow from them rests on lawyers who may be instructed to make allegations and on the court before which they can be made.' N. S. Paul, *Principles for the Press: a Digest of Press Council Decisions, 1953–84*, Press Council, London, 1985, p. 185.

crime, which is such an inhibiting factor in the lives of so many insecure or elderly people.

Following the Moors murders trial of Ian Brady and Myra Hindley at Chester Assizes in 1966, in which one of the leading witnesses admitted that he had received weekly payments from a newspaper under a contract that involved the provision of information, the Press Council made a Declaration of Principle after consultation with editors. This stated unequivocally that no payment or offer of payment should be made by a newspaper to any person who was, or was expected to be, a witness in criminal proceedings in exchange for any story or information connected with the proceedings until they had been concluded. To preclude the colouring of evidence to be given in court, it was also declared that no witness in committal proceedings should be questioned on behalf of a newspaper about the subject matter of his evidence until the trial was concluded. The Press Council's declaration did not confine itself to witnesses, going on to affirm:

No payment should be made for feature articles to persons engaged in crime or other notorious misbehaviour where the public interest does not warrant it; as the Council has previously declared, it deplores publication of personal articles of an unsavoury nature by persons who have been concerned in criminal acts or vicious conduct.[4]

In 1970 an ingenious editor failed to convince the Press Council that a payment for the memoirs of Ronald Biggs, the escaped train robber, fell outside the terms of the declaration because it had been made not to a person who had been convicted of a very serious crime, but into a trust fund for the benefit of his children. The Council refused to accept this argument, commenting: 'It is particularly harmful for the press to allow itself to be used as a vehicle for criminals who are under sentence but at large and thereby enable them to enrich their dependants or friends.'[5]

Four years later, the Committee on Contempt of Court returned to the question of offers made by the press to purchase stories from witnesses in legal proceedings, usually for publication after the trial, especially where the amount of the payment

---

[4] 'The Press and the People', *31st Annual Report of the Press Council 1984*, London, 1985, p. 232.
[5] A report by the Press Council, *Press Conduct in the Sutcliffe Case*, Press Council Booklet no. 7, London, 1983, p. 83.

was contingent on the outcome (as was alleged to have occurred in the Moors murders case). The Committee regarded the potential dangers as being 'sufficiently grave to warrant further inquiry as to its prevalence and, if found necessary, legislation to restrain or wholly prohibit this practice'.[6] The Royal Commission on the Press, which sat between 1974 and 1977, followed up this suggestion, investigating all of the matters covered by the Press Council's declaration.[7] No evidence was found of payments having been made in the intervening period to witnesses or potential witnesses of the kind that had been so strongly criticized earlier, but the Commission reported:

Some newspapers are prepared to buy stories from those associated with criminals, on the justification of 'public interest' which, in this context, must mean no more than that they judge the public will be interested to read stories serving only to excite the prurient or morbid curiosity of their readers. We believe that those who write, or lend their names to, these stories are under considerable pressure to exaggerate their most sensational features. This tendency is likely to be increased when different papers compete for the same story.[8]

The Royal Commission did not recommend legislation to deal with chequebook journalism, but urged the Press Council to ensure that its declaration was obeyed and to say when, in the Council's view, individual papers had breached the terms of the declaration. It was not long before another notorious trial put good intentions to the test. The arrest and trial of Peter Sutcliffe in 1981, after a police investigation lasting over the previous five years, provided a temptation that was irresistible. A total of 150 journalists from the British and foreign press attended the committal proceedings in Dewsbury Magistrates' court. No members of the public were present, although a crowd of about a thousand people gathered outside the court.[9] The next day, newspaper front pages and main pages were filled with accounts

[6] *Report of the Committee on Contempt of Court* (Cmnd. 5794), HMSO, London, 1984, p. 35.

[7] Royal Commission on the Press, *Final Report* (Cmnd. 6810), HMSO, London, 1977. This was the third Royal Commission on the press since 1947. The original chairman was a High Court Judge, Sir Morris Finer, who died in December 1974 and was succeeded by Professor O. R. McGregor. He was made a life peer as Lord McGregor of Durris in 1978.

[8] Ibid., p. 103.

[9] *Press Conduct in the Sutcliffe Case*, p. 9.

of the arrival of the accused man at the court, the brief proceedings that had taken place, and interviews with his wife and
members of his family. Not only was Sutcliffe named, but there
were pictures of him, his wife, and their home. What was not
apparent was that the first bids for his wife's exclusive story had
already been pushed under the door of the house in which she
was taking temporary refuge, and that Sutcliffe's father and other
members of his family were being accommodated as the 'guests'
of a newspaper at a hotel across the Pennines in the Lancashire
countryside.[10] By the time the case came to trial at the Old
Bailey, the Attorney General found it necessary to warn the jury
that some of the witnesses had been paid or offered sums of
money by the press.[11]

Adverse public and parliamentary reaction to the conduct of
the press included six petitions, each with between 100 and
1,000 signatures, which were forwarded to the Press Council by
MPs representing northern constituencies; critical motions in the
House of Commons; and a letter sent on behalf of the Queen to
the mother of one of the victims expressing the Queen's 'sense of
distaste' if substantial sums had been paid to members of the
Sutcliffe family for their stories.[12]

At a relatively early stage, after the initial proceedings in the
Magistrates' court and before the trial, the Press Council
announced its intention to hold an inquiry into the press
coverage of the events and the reaction to it. The subsequent
investigation was a thorough one, covering the provincial press
as well as national newspapers and magazines but excluding the
broadcast media, and it lasted for nearly two years. The main
issues, set out in a full report which was published in 1983, were
the degree to which press reporting might have prejudiced a fair
trial and the presumption of innocence of the accused; the
allegations of chequebook journalism and the payment of what
had been criticized as 'blood money' (offers or payments to
relatives and associates of a man subsequently convicted of
multiple murder); and the harassment by journalists of people
connected with the case. In a forthright report, the Council
censured four national papers by name—the *Sun*, the *Daily Star*,
the *Daily Mail*, and the *Daily Express*—for breaching the

---

[10] Ibid., p. 11.   [11] Ibid., p. 12.   [12] Ibid., p. 13.

Declaration of Principle, adding in a rider that it deplored an attempt made by the *Daily Express* to mislead the Council in its inquiries. The same criticism was made of the *Yorkshire Post*, which was also condemned for making an offer to Sutcliffe's wife at a time when she could reasonably have been expected to be called as a witness.[13] Although the *Sun* had breached the Declaration by making payment to a witness, 'a redeeming feature' was that certain information relevant to the case which came into its possession had been made available to the police.[14]

The Council was highly critical of the composite general impression conveyed by the press that Sutcliffe, even before he had been formally charged, was the killer of thirteen women and girls. It described the press coverage as having been 'inexcusably prejudicial and unfair'.[15] On harassment, the Council concluded that the relatives of those at the centre of the case, the victims, and the accused were all subjected to wholly unacceptable and unjustifiable pressures by journalists and other media representatives anxious either to photograph them or bid for the right to publish their stories. The targets of their attention, the report said, 'were people in deep personal grief or grave anxiety and they were harassed by the media ferociously and callously'.[16]

In the light of these findings, the Press Council published a further Declaration of Principle in 1983, extending its previous declaration covering 'persons engaged in crime or other notorious misbehaviour' to include payments made to the associates, family, friends, neighbours, and colleagues of criminals. In future, the Council intended to judge cases on these lines:

Just as it is wrong that the evildoer should benefit from his crime, so it is wrong that persons associated with the criminal should derive financial benefit from trading on that association.

What gives value to such stories and pictures is the link with criminal activity. In effect, the stories and pictures are sold on the back of crime. Associates include family, friends, neighbours and colleagues. Newspapers should not pay them, either directly or indirectly through agents, for such material and should not be party to publishing it if there is reason to believe payment has been made for it.

The practice is particularly abhorrent where the crime is one of

---

[13] *Press Conduct in the Sutcliffe Case*, p. 102–25.   [14] Ibid., p. 102
[15] Ibid., p. 75.   [16] Ibid., p. 162.

violence and payment involves callous disregard for the feelings of victims and their families.[17]

Peter Sutcliffe, the focus of what the Press Council referred to as a case which had gained unparalleled media coverage, was found guilty on 22 May 1981 of the murder of thirteen women and the attempted murder of seven others, having unsuccessfully pleaded diminished responsibility. He was sentenced to life imprisonment, the mandatory penalty for murder, with a recommendation by the trial judge that he should serve at least thirty years before release on licence was considered. His appeal against conviction was dismissed a year later by the Court of Appeal.

What can be said about the performance of the Press Council in the light of the conduct of the press in Sutcliffe's case and its aftermath? One of its most outspoken critics, the barrister and author, Geoffrey Robertson, although unconvinced by the method of investigation, considered the Council's published report to be the most impressive effort in its thirty-year history.[18] The Press Council is a private body with public functions, voluntary rather than statutory in origin, composed of an equal number of representatives of the press and members of the public, the latter selected by an independent appointments commission. It is presided over by a part-time legally qualified chairman of high reputation, the sequence since 1964 having been Lords Devlin, Pearce, and Shawcross and Sir Patrick Neill, QC. Sir Zelman Cowen, the chairman since 1983, is an academic lawyer who was Governor-General of Australia between 1977 and 1982. The Council has a small permanent staff and is funded largely by the newspaper publishers. It has never commanded much confidence among editors or working journalists (the National Union of Journalists withdrew from membership in 1980), whose alleged inaccuracies, misrepresentation, or other abuses it investigates.

But the Council is more than an agency for investigating complaints and pronouncing upon them. Unlike most self-regulatory bodies, it has no sanctions other than the impact of its

[17] Ibid., p. 193.
[18] G. Robertson, *People Against the Press: An Enquiry into the Press Council,* Quartet Books, London, 1983, p. 97.

adjudications and declarations upon public opinion and professional practice. It can neither fine, suspend, nor expel those who transgress. Little has changed since the Annual Report for 1979, in which the chairman of the Council wrote: 'I see no prospect whatever of universal consent being forthcoming from journalists, editors, proprietors and publishers for the Council to be enabled to impose penalties on a voluntary basis. For the conferment of such powers legislation would be essential.'[19]

Consequently it is not uncommon for editors who are censured to brazen it out, as happened with some of the newspapers which were so strongly criticized in the report on the Sutcliffe case.[20] For this reason the tone of the Council's statements on ethical standards, and the forcefulness and persuasiveness of the arguments deployed, interacting with the public response, is crucial if this device of voluntary regulation is to survive and avoid moves towards the even more unsatisfactory alternative of legal controls on what is published. The fact that legislative intervention is so alien to the traditions of a free press, however great the provocation, means that in practice the threat is regarded by most newspaper publishers and journalists as a remote one. Healthy as this situation may be, judged from the standpoint of the freedom to report and comment, it makes the Press Council's task in ensuring compliance with its declarations harder to perform. Trying to shame the shameless is bound to be an unrewarding pursuit.

Once convicted, except in the cases of a handful of the most notorious prisoners, curiosity in the fate of individual offenders declines rapidly. For the vast majority, their progress in prison and subsequent release passes unnoticed. Sometimes it is put to the Parole Board that early release may give rise to serious local concern, but this is usually very localized, and alternative release plans may be made to protect the prisoner from vengeance if it is thought that he is suitable for release on other grounds. The same applies to informants who may be at risk of retribution at the hands of those whom they helped to convict. More wide-

[19] 'The Press and the People', *26th Annual Report of the Press Council 1979*, London, 1982, pp. 1–2. Chairman's Foreword: 'A Personal View' by Patrick Neill, QC.

[20] The Editor of the *Daily Mail* was quoted as having described the report as 'short-term, short-sighted and smug', proving 'yet again that the Press Council still does not truly understand the concept of a Free Press'. Robertson, *People Against the Press*, p. 4.

spread reaction depends on continued exposure in the news media, which have the capacity to keep public anxiety alive over very many years. Lord Longford, who has made such a noteworthy contribution to penal policy, has closer knowledge than most of the plight of some well-known offenders who have maintained a high level of attention and interest. In a House of Lords debate in November 1986, he had this to say about public opinion and its effect on prisoners:

Can we all agree that public opinion in regard to prisoners is very much influenced by the tabloid press? Can we all agree that the attitude of the tabloid press towards prisoners is quite deplorable? Once we agree with that, the question arises: how far are we going to pander to an opinion whipped up in that way and exploited for commercial reasons?[21]

Longford continued in strong vein, giving examples and censuring attitudes that were 'totally without scruple' and 'unbelievably vicious'. He goes too far.in inferring that the fear of adverse publicity, or a 'pandering' to prejudicial opinions expressed in the popular press, overrides all else when it comes to decisions on release. The Parole Board will take account, as it always has done, of the gravity of the offence as well as the likelihood of re-offending when considering whether or not to recommend release of a prisoner on licence. A person convicted of a terrorist bombing or of a very serious crime of violence, such as somebody who throws acid in the face of a girl who has jilted him,[22] and who receives a lengthy determinate sentence as a result, may not be successful in obtaining parole on first application (or at all), even if his custodial behaviour has been good, his release plan is satisfactory, and there seems little prospect that he will re-offend. Wider considerations of public policy cannot be disregarded.[23]

[21] *Parl. Debates*, HL, 482 (5th ser.), col. 166, 18 Nov. 1986.

[22] In an earlier debate in the House of Lords, on an Unstarred Question by the Earl of Longford on parole, Lord Harris of Greenwich (a former Parole Board chairman) and Lord Donaldson of Kingsbridge (President of NACRO) cited these circumstances as examples of where public feelings should be taken into account. See *Parl. Debates* HL, 481 (5th ser.), cols. 1057–95 at c. 1076 and c. 1067, 4 Nov. 1986.

[23] That there had been no change in practice from one government to another was indicated in a comment made by the former Labour Minister, Lord Harris of Greenwich, from the SDP benches: 'We cannot totally pretend that public opinion does not exist. Of course, we must not take a cowardly view, being frightened of being attacked by the *Daily Express*, the *Daily Mail* or even the *Sun*; but it is right, I think, for us to have a decent regard for the anxieties of our fellow citizens.' Ibid., col. 1076.

Condemnation of the attitudes of the tabloid press towards convicted prisoners ought to be balanced with the contrasting approach to be found elsewhere in Fleet Street. Some parts of the quality press, notably the *Guardian* and the *Observer,* have taken a consistently informed and thoughtful interest in penal questions and prison reform. There are early signs that the *Independent* may be set to follow the same honourable course. Outside the news bulletins on television and radio, both the BBC and Independent Television have helped to define the context in a number of specially commissioned and responsible documentaries. Moderate opinions emanating from these sources have had a definite, although unquantifiable, effect on the climate of official opinion in which decisions are taken. Newspaper influence, on penal policy as elsewhere, is neither homogeneous, nor negligible, nor wholly harmful. The stridency, sensationalism, and bias of the tabloids are often objectionable, as the Press Council has made abundantly clear, not simply as a matter of taste, but in the way they bear on the workings of criminal justice. Against this, miscarriages of justice can be brought to light and officialdom held to account for its actions.

III

The formulation of opinion and its application in the setting of criminal justice are contained, however loosely, within the confines of a liberal and essentially libertarian tradition. This may seem a large claim to those who resent the influence of public opinion (irrespective of the source) as a crude and emotive intrusion, introducing an uncontrolled and irrational element, almost primeval in character, into what should be a rational and consistent process reflecting principles of fairness and certainty. But the fundamental notion on which the whole edifice depends is that the individual citizen has rights and freedoms that can be abrogated only in ways and to an extent that are authorized by Parliament. It is this simple idea that prevails at the heart of the liberal tradition. The citizen has a corresponding duty to obey the law which exists for the protection of himself and his fellow men. If he breaches this obligation, he will be punished proportionately to the gravity of his offending. Retribution in this sense is central to public policies towards crime. Overlaid are other objectives and justifications which have changed as time has gone on.

For much of the twentieth century, retribution was unfashionable as a justification for the imposition of punishment. But those days are over. The danger now is that the philosophy of just deserts is so strong that it casts its shadow over all the other justifications for punishment: the aim of reforming or rehabilitating the offender so that he may lead a better life in future, with the public being spared the consequences of his repeated offending; the possibility of deterring some others from following down the same road; and the expression of public abhorrence of the crime by symbolic denunciation. All of these have a place, being overlapping rather than exclusive, the last in particular relating directly to the need to maintain public confidence in the system of criminal justice. To a lesser extent, it can also be regarded as an emotional satisfaction of victims' grievances against those who have done them harm.

Part of the attack on the rehabilitative ideal, touched on more than once in the preceding chapters, is that it seeks to impose on a free spirit a preconceived picture of what the socially healthy citizen ought to be and of how he should comport himself. Some theorists object to what they see as a form of re-education on the grounds that when an offender commits a crime he should sacrifice only those rights that are proportionate to the harm he has done. Thus the state may say to the citizen you must not steal, and if you do you may be punished by the imposition of penalties sanctioned by Parliament; but it may not attempt to compel respect for private property. More prosaically, it is argued, such attempts as are made to help prisoners improve their characters within the inauspicious and highly artificial prison setting, where they are cut off from family and income and are associating entirely with others in trouble, are unlikely to lead to much in the way of successes. There can be few tools of social engineering less promising than a prison sentence. As the high hopes have faded, undermined by recidivism and questioned by research findings, scepticism has given way to a growing mood of resignation, a feeling that nothing that has been tried has worked and that all that can be done is to ensure a reasonable standard of secure and humane containment in custody. Arguable as this outlook is, it must be prevented from escalating into a paralysing nihilism. The search for effects, for proof of rehabilitation, is natural enough, but it is one that can be destructive if taken to the extreme of repudiating the validity of any ideals implying worth, desirability, or belief in redemption, unless and until their usefulness can be proved.

That said, consequences do matter, and no penal theory could survive for long if it ignored them. So before writing off rehabilitation as a discredited ideal from a bygone era, and dismissing the efforts made by many well-meaning people to help offenders to stop offending and lead a better life, let us come down to earth and review, for the last time, some factual case histories. Unlike most of the examples cited earlier in the book, which contained very little to uplift the spirit, the case studies that follow have been selected precisely because they do hold out some hope.[24]

First of all there is the experience of a former caretaker (FC), aged forty-one when he came to a probation hostel in south London. Only three years before, he had been living with his wife and children and had had a full-time job as a local authority caretaker employed on a council estate. He was a good-hearted and patient man, inclined to take the cares of the estate on his shoulders. Residents called in with their problems at all hours of the day and night, and this put a strain on his marriage. Domestic pressures built up until finally his wife left him, taking the three young children with her, and he lost his job. A downward slide of sleeping rough or in night shelters followed, combined with heavy drinking. This way of life in turn led to petty crime and a spell in prison. After release, the FC went first to a large institutional hostel, moving on to a smaller hostel originally managed by the probation service but then taken over by NACRO. In a variety of different housing projects NACRO provides places for over 450 offenders and other single homeless people, either in hostels or in units of shared accommodation.[25] Of these, about

---

[24] Details supplied by NACRO from its own records and a case study reported by the North London Education Project. Although each of the examples given is London-based, provision for the resettlement for ex-offenders exists in other parts of the country.

[25] NACRO, *Annual Report 1985–86*, p. 12. In 1986 NACRO reported a total of 196 projects at which, on any day, around 20,000 people were receiving help in housing, education, employment training or crime prevention. Ibid., p. 11.

200 places in 1986 were in London and 250 in the provinces. The emphasis is on learning to cope with life outside institutions and trying to find longer-term solutions to the enduring problems faced by the homeless ex-offender.

The co-ordinated accommodation scheme (CAS) provides supported housing for single homeless people in inner London. It was set up by NACRO and the probation service in 1984 and by March 1986 had 115 places in sixteen houses in different parts of London.[26] The houses are shared by a cross-section of homeless people, mainly in the eighteen to twenty-five age range and not all of them ex-offenders, although those who have previously been through the courts form the largest group. In spite of being a good deal older than most of the other residents, the FC fitted in well. He did not re-offend, and he learned to control his drinking. He was nominated to the local council for rehousing and some months later was successful in obtaining a self-contained bed-sitter. This he decorated to a high standard despite the very limited funds available. Soon after he was able to obtain part-time clerical work which offered a good possibility of full-time work in the future. He kept in touch with the staff of the CAS scheme, whom he regards as his friends, visited them occasionally, and seemed to be settled and contented in his new life.

The second example is from an education project in north London which was launched in 1980 to provide education as well as accommodation and support for ex-offenders of either sex. The project was seen by its founders as being a joint response to the combined disadvantages of homelessness and poor, or unfinished, education.[27] Education can lay down a constructive path towards resettlement, but a helping hand is needed to assist the would-be student through what can seem a bewildering and frightening maze. The North London Education Project is a broad one, ranging from basic literacy and numeracy, through vocational training courses to O and A levels, and ultimately to university degrees. Thirty residential places are provided by housing associations in Hackney and Islington, each resident being allocated a study bedroom and sharing a kitchen and bathroom. Non-residential provision exists at a day centre,

[26] CAS, *Annual Report 1985–86*, p. 5.
[27] North London Education Project, *Report 1984–86*, p. 2.

which is not limited solely to offenders, and through the North London Further Education College, where one of the Project's education officers is based.

The potential of education as an instrument of rehabilitation was mentioned in the earlier chapter on prisons. Sometimes an offender becomes interested in study while in prison, having been put off by unrewarding school experiences, and starts climbing a ladder of educational achievement that continues after release. One young woman had experienced an upbringing which resulted in a severe lack of self-confidence and social skills, and was able to find acceptance only in a peer group involved in prostitution and other criminal activities. She eventually received a custodial sentence, and while in prison became fired with an enthusiasm for studying and education that grew over time. Starting from a low level of literacy, and with the help of the prison education department, the determined student (DS) took and passed a Pitman's Elementary English Language course, followed later in the year by the RSA English Language Stage 1. She became set on pursuing her studies on release and applied for a residential place at the North London Education Project. There she studied English and mathematics for the rest of the academic year, moving on to an O-level course in the following year. The DS worked hard and obtained three O-level passes, in English language, sociology, and psychology. That summer she went to Israel and worked on a kibbutz for four months before returning to take A-levels in English literature and law. Once again she was successful, passing in both subjects. After two-and-a-half years she was rehoused by a local housing trust which co-operated with the educational project. Her scholastic career continues. She has entered for an A-level philosophy course and an RSA typing qualification, and is working part-time for the Citizens' Advice Bureau, having gained by her own efforts a measure of personal insight and a sense of self-worth that were so critically absent in her earlier career.

The subject of the third case had most of the features of the classic recidivist (CR). By the age of thirty he had accumulated twenty previous convictions and had served several prison sentences. In applying for parole from a twenty-seven-month sentence for offences of theft and assault, the CR was able to show that he had obtained a place at the 134 Project in south

London. This resettlement scheme, aimed at providing housing for men discharged from prison in the London area without a home address, was opened in 1972 as a result of collaboration between NACRO and the Inner London Probation Service's after-care unit. It had originally been thought that a ten-week stay would be long enough to enable most of the residents to make the adjustment to living outside the prison and finding something to do and somewhere to live. It had soon become apparent, however, that longer-stay accommodation was also required, and so a distinction was made between intake houses, where residents would stay during the first few weeks for acclimatization and assessment of immediate and longer-term needs, and second-stage houses, to which residents could be moved when and if places were available. From the start, the aim was to encourage residents to do things for themselves, rather than have the staff do things for them, within as non-institutional an atmosphere as could be created. The prerequisites of resettlement were seen as secure accommodation and the acquisition of some of the skills and information needed to cope with the stresses of independent living. This is particularly relevant for those, like the CR, who had spent so large a proportion of their lives in institutions.

The work with residents revolves around the pressures they must expect to meet on leaving their more sheltered environment and re-entering a world where the standing temptation is to see re-offending as the easy option. In the case of the CR, it was his temperament that had led him into trouble so often before. When things went wrong he became frustrated and was knocked off-balance, his aggressiveness being released by drinking. His stay at the project lasted well beyond the six months on parole, amounting to thirteen months in all. He lived initially at an intake house, then moved on to second-stage housing before finding a flat of his own through one of the housing associations linked with the project in south-east London. During this period the CR responded positively to the support he was receiving, only once getting into trouble when he was charged with an assault on a policeman after a New Year's party. For this offence he was fined, but he paid off the fine and gave up alcohol. In his private life, he survived the emotional shock of his mother's death and became engaged to be married. In December

1986, living at his own flat, still maintaining contact with the 134 Project, and with the prospect of a permanent job and married life in front of him, the CR had been out of prison for nearly two-and-a-half years. It may not seem a long time to those whose lives have run a different course; but the chances for the future, for the public as well as for a man with such a dismal past, must be brighter than if no attempt had been made towards his rehabilitation.

IV

Whatever the doubts of theoreticians, judges, or statisticians about the ability of rehabilitation to reduce the overall incidence and repetitive quality of criminal offending, it is hard to ignore the implications of case histories such as these. While the successes may not be any more typical than the failures, they are not isolated examples. In any scale of values, it must be better to have somewhere to live than to be homeless; to have some friends and supporters than to be friendless and alone. Moreover, the act of trying to help offenders is laudable in itself, quite apart from the results. It is curious and rather depressing how the idea of treatment or welfare has come under attack simultaneously from left and right as a dated and inappropriate approach to social policy. To some it is seen as intervention that is patronizing and manipulative, an attempt by well-meaning outsiders to impose solutions drawn from their own very different lives, rather than centred on the experiences and problems faced by those with whom they are dealing. Such endeavours, it is claimed, tend to produce arrogance in the treater leading to a form of injustice if the powers assumed over the life of the offender are greater than can be justified by his crime.[28] Others, at the opposing end of the political spectrum, are to be found proclaiming that welfare is in everyone's worst interests[29] because it raises false hopes which experience has

[28] A. E. Bottoms and W. McWilliams, 'A Treatment Paradigm for Probation Practice', *British Journal of Social Work* (1979), 159–202.

[29] P. Morgan in C. Brewer, T. Morris, P. Morgan, and M. North, *Criminal Welfare on Trial*, Social Affairs Unit, London, 1981, pp. 58–9.

shown are incapable of fulfilment, failing to protect the citizen and subverting the rule of law.[30]

Dialogue on these lines interests expert opinion more than it does the general public. There is little evidence to indicate that public attitudes, as distinct from the ideas of the small group of people who take an enlightened interest in penal reform, ever moved very far towards the rehabilitation of the offender. It was an ideal adopted, practised, and finally questioned by those working closely with offenders or having a voice in framing or carrying out criminal justice policies. For many years, before disillusion set in, rehabilitation provided a guiding light which fortunately is not yet extinguished. In the hostels, the day centres, and probation offices, statutory and voluntary workers are still inspired by the desire to help their fellow men, although few expect that their support and advice will lead to the transformation of characteristics that may have been hardened by long years of deviant behaviour and institutional confinement.

The dichotomy between expert opinion, in the sense of the views of an informed élite, and popular opinion, whether expressed or anticipated, runs throughout the debate on crime and what to do about it. Although no one can claim to speak with any precision, the public's concern seems steadfastly to be centred upon the need for society to be protected from the offender. Beliefs in retribution or rehabilitation are seldom articulated unless drawn out by leading questions in sample surveys and opinion polls. The reliability of polls as accurate indicators of public attitudes is suspect, depending on the size and representative nature of the samples, the phrasing and content of the questions, and the context in which they are put. Despite these reservations, which apply less to the British Crime Survey than to opinion polls on crime and punishment commissioned by newspapers, often at short notice and sometimes with ulterior motives, sample surveys can provide pointers to public attitudes which relate to policy formation. The clearest indication to have emerged to date is that, on the basis of the most reliable information available, public opinion is less punitive regarding crime and how it should be dealt with than is

[30] Ibid. Preface by Digby Anderson, series editor, p. 6.

often assumed.[31] The British public as a whole is not portrayed as being particularly bloodthirsty or vindictive. The British Crime Survey's conclusions are summarized as follows:

Certainly the findings are at odds with the impression which opinion polls tend to give of a thoroughly punitive public. If people are asked at a general level whether court sentences are adequate, a great majority answer that they are not. But if they are asked—as in the BCS—about a specific incident involving themselves, and thus have a concrete example upon which to base their judgments, they are less punitive.[32]

The difference in attitude between those with first-hand knowledge of a crime and those without was shown in the finding that, whereas only 36% of victims of burglary thought that imprisonment was warranted for 'their' offender, some 62% of the general public favoured it for a man of twenty-five with previous convictions who was sentenced for burglary.[33] The actual statistic of prison sentences imposed in 1983 on men aged twenty-five (including first offenders) who had been convicted of burglary was 61%.[34] Court practice thus appeared to be broadly in step with the attitudes of the public towards sentencing levels in general, but there was a marked tendency to overestimate the leniency of the courts, which helped to explain why some polls disclosed a popular demand for tougher sentencing. Where there is first-hand knowledge of offences, public attitudes can be even more lenient than those of the courts.

In the case of residential burglaries, at least, there are no empirical grounds for the courts to believe that the public is looking for more punitive sentencing. There is solid backing for non-custodial penalties for non-violent offenders. Where crimes have led to violence to the person, however, public attitudes have become increasingly hostile, suggesting a collective feeling that sentences for property offences and those entailing violence have got out of kilter. This response encourages the claim periodically

[31] The relevant research is summarized in a NACRO briefing paper on *Public Opinion and Sentencing*, published in November 1986.

[32] M. Hough and P. Mayhew, *The British Crime Survey: First Report* (Home Office Research Study no. 76), HMSO, London, 1983, p. 28.

[33] Perhaps surprisingly, no more than 2% of victims proposed judicial corporal punishment when asked what treatment the offender deserved to receive. Ibid., p. 28.

[34] M. Hough and P. Mayhew, *Taking Account of Crime* (Home Office Research Study no. 85), HMSO, London, 1985, pp. 43–4.

made by the courts that the penalties imposed have a symbolic function, expressing and reinforcing the extent of public disapproval of the offence. Out of a multiplicity of judicial pronouncements from the Bench, we can select some remarks by one of the most experienced of all criminal judges, Lord Justice Lawton:[35]

> society through the courts must show its abhorrence of particular types of crime, and the only way in which the courts can show this is by the sentences they pass. The courts do not have to reflect public opinion. On the other hand courts must not disregard it. Perhaps the main duty of the court is to lead public opinion. Anyone who surveys the criminal scene at the present time must be alive to the appalling problem of violence. Society, we are satisfied, expects the courts to deal with violence ... The time has come, in the opinion of this Court, when those who indulge in the kind of violence with which we are concerned in this case must expect a custodial sentence.[36]

Seen in this way, the courts stand in a dual relationship to public opinion, influencing it by expressive sentencing and responding to it. There are differing considerations to be reconciled, and the Judicial Studies Board, referred to in the opening chapter, provides the judiciary with a convenient forum to assist in their resolution. Informed opinion has more of a one-way quality, publicizing, campaigning, and seeking to persuade a wider public towards particular points of view. Governments and others responsible for the formulation of criminal policy need to take note of both varieties of opinion. Expert opinion (although seldom unanimous) can be canvassed by way of green papers and assessed via direct contacts. General public opinion is shifting and hard to identify. Its role tends to be residual, an inhibiting factor more often than a positive one.

Politicians pride themselves on their ability to take the public pulse and to respond to what it tells them. Just how this is done is one of the arcane mysteries of political life. Anecdotal experiences, letters from constituents and interest groups, questions put and opinions expressed at public meetings, merge

---

[35] Sir Frederick Lawton, son of a prison officer who became a prison governor, was a judge in the Queen's Bench Division of the High Court of Justice 1961–72, and a Lord Justice of Appeal 1972–86. He also served on the Criminal Law Revision Committee from its formation in 1959 and as chairman 1977–86.

[36] *R. v. Sergeant* (1974) 60 Cr. App. R. 74 at p. 77.

with the powerful currents of press reporting and comment to produce an impressionistic picture in the mind of the elected representative. Since most of those who have been elected wish to be re-elected, they have a compulsion not shared by officials, pressure groupers, journalists, or other interpreters of the public mood. Penal policies are adopted, amended, or abandoned because of what is regarded as public concern, whether expressed or anticipated. Policy-makers, officials as well as ministers and parliamentarians, are constantly aware of the restraining hand of public attitudes and opinion. External influences must be reconciled with considerations of cost and practicability, as well as with principles of fairness and consistency. Yet the capacity of public opinion to get in the way of the best intentions of the well informed is a constant lament of officials and special interest groups. The fact that populist attitudes do not always carry the day is seen in Parliament's reluctance to accept the reintroduction of the death penalty for murder. That, however, is a rare example, although a most important one, of a deliberate policy maintained by a majority of elected MPs on a non-party basis, in defiance of what most see as the will of a substantial (and highly vocal) part of the electorate.

Dr Andrew Ashworth has postulated élitist and populist models of penal policy-making.[37] An élitist model would require decisions to be taken by bodies insulated from direct political pressure, the justification being that penal policy is a serious social issue requiring an informed and rounded approach which can be expected only of experts such as judges, prison administrators, police, those concerned with victims' support, or others with claims to sufficient knowledge. Discussion and decision-making would be in terms of a public interest objectively defined, rather than by reference to majority opinions expressed by members of the public. The structure might maintain some separation between the legislative, judicial, and executive spheres, or it might attempt some redistribution, for example by the creation of an independent sentencing commission. A populist model of decision-taking, in contrast, would be led by public concerns to which politicians and

---

[37] *Ditchley Conference Report*, no. 13, 1985/6, pp. 4–5. Dr Ashworth is editor of the *Criminal Law Review* and a Fellow of Worcester College, Oxford.

administrators would be expected to respond. These could be channelled through a network of elected community councils to which penal decisions might be delegated. In any populist model, the press and other media of mass communication would wield considerable influence. Depending on the popular will, the results might be longer sentences and higher imprisonment rates, together with capital punishment or, conversely, what might be regarded as non-repressive measures of prevention, restitution, and supervision in the community. The methods of identifying and measuring opinion would be crucial, and much would depend on public presentation and access to the mass media.

After making his spirited and imaginative analysis, Ashworth accepts that neither model matches the actuality of the complex society in which we live. Elements of an élitist model survive in some of the more sheltered corners of the penal system where, until recently, the penetrating beam of publicity has not shone brightly. Examples of low-visibility procedures are prosecution practices and parole. In Britain, still some way behind the United States but subject to the same pressures, there is demand for greater responsiveness throughout the criminal justice system. The heightened emphasis on victims' support and victims' rights can to a degree be attributed to populist forces, and the same may be true of neighbourhood watch and some other preventive initiatives. In our pragmatic British way, it is unlikely that we shall move very far towards either popular or élitist extremes. The two strands will continue to mingle in the way policies are formulated and implemented, sometimes coinciding and some-times in collision, but on the whole constituting a reasonably representative basis for actions by agencies of the state which bear so profoundly on the lives of many of its less fortunate citizens.

V

There is an unreality in public expectations as to what can be achieved by any process of criminal justice. Laws and their enforcement, the detection and prosecution of offenders, their trial and sentence in the courts, the administration of the prisons and non-custodial alternatives—all form part of what can loosely be called penal policy. But they do not constitute a systematic and unified whole that can readily be held out for inspection or approval. Where there is so little agreement as to what the objectives of policy should be, it is hard for administrators to assess how they can best be delivered in terms of cost and effectiveness. The truth is that we have inherited a piecemeal set of policies and practices which can be understood only by description. That is the main justification for the mechanistic approach adopted in this book. It describes how the main penal policies work in practice—their strengths and flaws—and the way in which new, and sometimes unrelated, developments are growing up alongside them. Together they amount to a check on offending, a way of punishing the guilty and reducing to the greatest extent possible the risk to members of the public of being victimized by crime.

In the end, it will be factors going far beyond the reach of governments, the courts, and the penal system that will determine the level of criminality in the future. Demographic changes are likely to prove especially significant. The next decade will see a massive decline in the number of young males aged between ten and twenty. The concentration of offending in the younger male age group has been explained in an earlier chapter, and the size of the fourteen to sixteen group is already in sharp decline. The age group from seventeen to twenty is set to shrink from 1.7 million to 1.2 million between 1986 and 1995.[38] Half a million fewer young men should mean fewer criminal offences, other things being equal. It does not necessarily follow, however, that other things will be equal. Fewer, but nastier, young men could do as much, or even greater, damage to society.

It should not be expected that a reduction in the number of young offenders passing through the criminal justice system will automatically lead to a corresponding reduction in the number of custodial receptions. This is because young men and boys are less likely to be sentenced to imprisonment than are adult offenders. Since prison should be used only as a last resort, sentencing courts are properly reluctant to impose custodial penalties on younger people convicted of less serious property offences until they have built up a record of previous

[38] D. Riley, 'Demographic Changes and the Criminal Justice System', Home Office, *Research Bulletin*, 20, 1986, p. 30.

convictions. In addition, the overall size of the prison population is determined by the length of sentences as well as by the numbers sentenced. There is little to suggest that the trend towards longer sentences will be reversed in the immediate future, and certainly not for those convicted of violent offences. The numbers on remand awaiting trial or sentence also show no signs of abating. Unless, as is proposed in Chapter 8, remand prisoners can be got out of the prison system altogether, the probability is that these considerations will combine to deny the prisons the relief they so urgently need until the new places currently under construction come into use. Important though the demographic effects are, it would be foolhardy to rely on them as a reason for halting or slowing down the prison-building programme.

What can we expect from altered attitudes and patterns of behaviour? The generally permissive outlook that prevailed in the 1960s is now in retreat. Self-expression was the agent that loosened the grip of the conventional social controls of the family, religion, and the myriad forms of established authority, stuffy and restrictive as they appeared to be to many young people. Permissiveness, if we use the word as a token, encouraged experimentation of all kinds, tolerated promiscuity and deviant sexual relationships, and allowed the use of hard drugs to spread insidiously from small groups until by the 1980s it had become widely established. The damage done by drug abuse is now only too evident, and the growth of an extensive and illegal service organization, promoting the use of drugs as well as supplying them and costing the state enormous sums in enforcement, is a continuing incentive to crime.

If attitudes are changing, as it seems they are, there will be other dangers, self-deception and authoritarianism among them. We should beware of thinking that when attitudes change behaviour follows, for it need not be so, or at any rate the two do not always ride in tandem. We need look no further than drink–driving for a contemporary example of the gulf between beliefs and action. Contradictions abound; it is in the nature of man that he sometimes does what he does not believe in. Moreover, crime, even the most serious of crimes, including homicide, may result from a lack of foresight rather than deliberate intent. Nor is it axiomatic that more authoritarian and less permissive attitudes would necessarily lead to a reduction in crime. Unless carefully managed, too abrupt a transition could result in an alienated minority among whom crime might be even more prevalent than at present. But attitudes are the starting point, the precursors of changed behaviour. That is why the educational campaigns on drug and alcohol abuse are so important, why they must be maintained, and why they hold out some grounds for hope.

The prospect of an AIDS epidemic provides an opportunity as well as a most ominous threat. To meet it with fear and intolerance, ostracizing homosexual and heterosexual carriers, actual or suspected, would be the worst of all responses. Calculated self-interest may be expected to lead to some reduction in the frequency of casual sexual intercourse with various partners and a greater use of condoms as a barrier to AIDS infection. That is a practical course of great importance in terms of public health and personal safety. But more profound, and ultimately more effective as an antidote than any physical forms of protection, would be the acceptance of a very different code of morality to that which allowed the epidemic to take root. It is fruitless, and irrelevant, to speak of a return to traditional moral values. The moral values of the past are open to conflicting interpretations and do not in any event provide a ready-made suit of armour likely to fit comfortably the citizen of today. In the case of AIDS, self-interest, always the most potent of motivators, should be accompanied by compassion for those who are suffering, some of them fatally, from a disease for which there is as yet no known cure. Sensitivity towards others and self-restraint are values that would not only help to check the spread of a terrible disease, but could expand to effect behaviour more widely. It is not unknown for good to come from adversity, painful though the process may be. But it takes vision and courage to recognize and make the most of the opportunities that beckon in the gathering gloom.

All this may seem a far cry from offenders and offending, and it is time to finish. Throughout the book, the message comes through that criminals are more often made than born, and that the conditions in which they are brought up, or are treated by the penal system when they have transgressed, can have a lasting effect on subsequent conduct. Human relationships are the key— love, example, and encouragement; consistency and fairness in punishment as in rewards—whether the setting is the private home or the public institution. No one offender is the same as another, and no outsider can ever be sure just what it was that led a man or woman to do what they did. Sometimes the crime is one that is unimaginable to most people, but in many other incidents there is a strong element of 'there, but for the grace of God, go I'. Motives and circumstances are boundless; the responses deployed by society should be similarly elastic. Common humanity is the moral sieve through which criminal policies aimed at protecting the public must pass if they are to be regarded as deserving the title 'justice'.

Index